"VULGAR AND MECHANICK"

The Scientific Instrument Trade in Ireland
1650-1921

ROYAL DUBLIN SOCIETY
HISTORICAL STUDIES IN IRISH SCIENCE AND TECHNOLOGY

Number 1
RICHARD GRIFFITH 1784-1878
Edited by Gordon Herries Davies & Charles Mollan
*1980, pp. vi + 221, ISBN 0 86027 005, IR£10.00**

Number 2
REPRINTED PAPERS ON BEE HUSBANDRY
Edited by Charles Mollan
*1980, pp. iv + 42, ISBN 0 86027 007 6, IR£3.00**

Number 3
JOHN TYNDALL – ESSAYS ON A NATURAL PHILOSOPHER
Edited by William Brock, Norman McMillan & Charles Mollan
*1981, pp. xii + 219, ISBN 0 86027 008 4, IR£10.00**

Number 4
SHEETS OF MANY COLOURS – THE MAPPING OF IRELAND'S ROCKS, 1750-1890
By Gordon Herries Davies
*1983, pp. xiv + 242, ISBN 86027 014 9, IR£15.00**

Number 5
FLOREAT HIBERNIA – BIO-BIBLIOGRAPHY OF ROBERT LLOYD PRAEGER
By Timothy Collins
*1985, pp. xiv + 141, ISBN 0 86027 017 3, IR£12.00**

Number 6
NOSTRI PLENA LABORIS – AN AUTHOR INDEX TO THE RDS SCIENTIFIC JOURNALS, 1800-1985
By Charles Mollan
*1987, pp. 120, ISBN 0 86027 018 1, IR£25.00**

Number 7
DUNSINK OBSERVATORY, 1785-1985 – A BICENTENNIAL HISTORY
By Patrick Wayman
*1987, pp. xiv + 353, ISBN 0 86027 020 3, IR£25.00**

Number 8
VULGAR AND MECHANICK – THE SCIENTIFIC INSTRUMENT TRADE IN IRELAND 1650-1921
By John Burnett & Alison Morrison-Low
*1989, pp. x + 166, ISBN 0 86027 026 2, IR£15.00**

ALSO AVAILABLE

IRISH NATIONAL INVENTORY OF HISTORIC SCIENTIFIC INSTRUMENTS – INTERIM REPORT 1989
By Charles Mollan
*1989, pp. 224, ISBN 0 86027 025 4, IR£25.00**

*1989 prices, subject to change without notice.
Order from: Dr R. Charles Mollan, Science & Arts Department,
Royal Dublin Society, Ballsbridge, Dublin 4.
Please add £1.50 for postage & package.
Cheques etc. should be made payable to the Royal Dublin Society.

ROYAL DUBLIN SOCIETY
HISTORICAL STUDIES IN IRISH SCIENCE AND TECHNOLOGY
NUMBER EIGHT

"VULGAR AND MECHANICK"

The Scientific Instrument Trade in Ireland 1650-1921

J.E. BURNETT and A.D. MORRISON-LOW

N·M·S
NATIONAL MUSEUMS
OF SCOTLAND

ROYAL DUBLIN SOCIETY

1989

ISBN 0 86027 026 2

Typeset by Wendy A. Commins; make-up by Paul Bray.
Printed in the Republic of Ireland by Mount Salus Press Limited, Dublin.

FOREWORD

It is only in the past few years that the History of Scientific Endeavour in Ireland has been studied to any extent. One reason, perhaps, for this was that the secondary sources which can be of such help to the historian – biographical dictionaries, bibliographies, directories and so on – have only recently started to appear. This volume, which looks in a detailed way at the scientific instrument trade in Ireland, is a further contribution. It is hoped that it will stimulate work in a new field which is proving considerably richer than some might have imagined.

It was a joint initiative between the Royal Dublin Society and the Royal Irish Academy which established the Committee on Historic Scientific Instruments in 1985, under the Chairmanship of Professor Gordon Herries Davies of Trinity College and the Secretaryship of Dr Charles Mollan of the Royal Dublin Society. Three Honorary Members from Great Britain were asked to advise – Dr J.A. Bennett of Cambridge University, Dr G. L'E. Turner, then of Oxford University, and Dr R.G.W. Anderson at the National Museums of Scotland in Edinburgh. This Committee has acted as a further boost to the study of the scientific instruments used by researchers, specialised practitioners and ordinary workmen, and has engendered a feeling of co-operation between Irish and British scholars. Dr Mollan, whose own institution has been so important in the development of science in Ireland, is now engaged in compiling a detailed inventory of historical instruments which should help both to ensure their long-term survival and to develop an appreciation of their value as evidence for the writing of history. Dr Mollan has already been able to publish an interim inventory which is the first real indication of the range and scope of the material which remains in universities, schools, societies and other institutions.

The authors of this volume are both based at the National Museums of Scotland. Their work follows in the tradition set by E.G.R. Taylor, who in 1954 and 1966 published volumes on British Scientific Instrument makers. In 1972 D.J. Bryden, also of the National Museums of Scotland, published a volume which dealt specifically with Scottish craftsmen. The efforts of John Burnett and Alison Morrison-Low have borne considerable fruit and it is most pleasing that our institutions are jointly publishing this contribution to the history of Ireland and the history of science.

DR R.G.W. ANDERSON

Director,
National Museums of Scotland

FOREWORD

When Dr Robert Anderson visited Ireland in 1982, he alerted us to the rich cultural heritage we possessed in our historic scientific instruments. Through personal contacts, he had been put in touch with the Royal Dublin Society and we had been able to arrange for him to visit some of our holdings. He was later joined in a second visit by Professor Gerard Turner, and they impressed on us the national and international importance of our collections. A joint meeting in Dublin between the Irish National Committee for the History and Philosophy of Science and the British Society for the History of Science, when Professor Turner was President of that latter body, gave a further boost. This enthusiasm led to the founding of the Royal Dublin Society/ Royal Irish Academy Joint Committee on Historic Scientific Instruments – the first Joint Committee between the two bodies since that set up in 1900 to study the eclipse of the Sun as seen at Plascencia in Spain. Dr Jim Bennett, originally from Belfast, Professor Turner and Dr Anderson agreed to act as Corresponding Members of the Joint Committee and they have freely offered advice from their rich store of knowledge and experience.

Dr Anderson has had a further important role. In his position as Director of the National Museums of Scotland he has been able and willing to encourage the work of two of his staff, the authors of this book, as they studied the Irish scientific instrument trade. It was he who suggested that the volume might be co-published, thereby further helping fruitful co-operation between our institutions.

Vulgar and Mechanick presents really new work, of the greatest importance to the study of Irish instruments. While a few publications have referred to the Irish instrument making trade, there was no solid and comprehensive work on the subject until now. As a result, when instruments with Irish signatures turned up (and they have been turning up in ever-increasing numbers in recent years), there was no literature available about many of the makers, their working dates, their range of products, their business and families.

I am glad that the authors have not fallen into the too-frequent trap of waiting until work is complete in every detail before publication. New makers are being discovered, and new information found, fairly frequently, and this will undoubtedly be incorporated in future publications. As the authors say, "we are acutely aware that this should be regarded as an interim report, because neither the questions we raised nor the sources examined are exhausted".

In a personal comment, I should like to thank John Burnett and Alison Morrison-Low warmly for their friendly generosity in allowing me to make use of their work while it was being compiled. I hope that they will continue their interest in Irish instruments and build further on the work which they have started. Undoubtedly this will generate wide interest in this fascinating part of our scientific, technical and cultural heritage. This joint publication, and the work which it records, are examples of very happy co-operation between Scotland and Ireland.

DR R. CHARLES MOLLAN

Science & Arts Officer,
Royal Dublin Society

CONTENTS

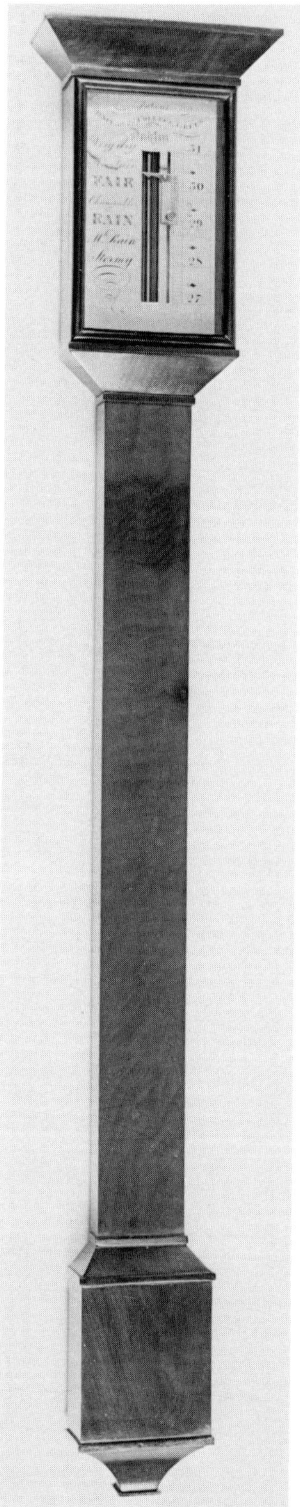

Frontispiece: Patent barometer, signed by Richard Spear. Patrick Marney.

ACKNOWLEDGEMENTS

This publication would never have seen the light of day without the goodwill and assistance of a wide range of individuals and institutions. Firstly, we acknowledge the generosity of the Royal Society of London for providing a grant to Alison Morrison-Low for travel and research in Ireland in 1987; secondly, we are grateful to both the Royal Dublin Society and the National Museums of Scotland for their continuing support throughout the project. Various institutions and archives have thrown open the doors to their collections and given us help, advice and information, and it is a pleasure to thank them here: in Belfast, the Linenhall Library, the Belfast Public Libraries, the Ulster Museum, the Ulster Folk and Transport Museum, and the Department of Irish Studies, Queen's University; in Cambridge, the Whipple Museum of the History of Science; in Cork, University College; in Dublin, the National Library of Ireland, the National Museum of Ireland, the Royal Dublin Society, the Royal Irish Academy, Trinity College and University College; in Maynooth, St Patrick's College; in London, the Royal Astronomical Society, the Science Museum, the Science Museum Library; and in Oxford, the Museum for the History of Science.

At a more personal level, we would like to thank Dr J.A. Bennett, D.J. Bryden and Dr R.C. Mollan for their comments on the whole text at manuscript stage; and Professor Patrick A. Wayman for his advice on Chapter Six. Dr Mollan in addition made available to us the results of, and photographs from, his on-going work on the compiling of an inventory of historic scientific instruments in Ireland. We would also like to thank the following individuals who have assisted on specific points: Dr R.G.W. Anderson; Jon Baddeley; Dr John Butler; Rev. Professor Michael T. Casey; Dr John Chaldecott; Jeremy Collins; Dr Maurice Craig; the late Michael Crawforth and Mrs Diana Crawforth of Project Simon; Dr Jon Darius; H. Dawes; Dr Alun C. Davies; Peter Delehar; Paul and Edith Egestorff; Brian Gee; Barbara and Michael Gray; Dr John Green; Dr Willem Hackmann; Professor Gordon Herries Davies; Peter Hingley; Professor Ian G. Hughes; Dr Anita McConnell; Dr W.F.C. Purser; Dr Alison Sheridan; Dr A.D.C. Simpson; Ken Smith; Miss M. Stokes; Martin Suggett; Dr John Teahan; Anthony J. Turner; Professor Gerard L'E. Turner; Dr Brian Walker; Natasha Weyer-Brown. Our gratitude goes to Mrs Kathleen Davidson, who typed endless drafts without complaint and corrected our spelling and grammar. Finally, our special thanks go to Roger Willbourn for accommodation in Dublin, and to Cathy Wilson for her welcome in Belfast.

March 1989

ALISON MORRISON-LOW
JOHN BURNETT

CHAPTER ONE

VULGAR AND MECHANICK: SCIENTIFIC INSTRUMENTS IN IRELAND

A.D. Morrison-Low

Their Houses are very ill built, the Walls bevil, without one right Angle in any Apartment; and this Defect ariseth from the Contempt they bear for practical Geometry; which they despise as vulgar and mechanick, those Instructions they give being too refined for the Intellectuals of their Workmen; which occasions perpetual Mistakes.

Jonathan Swift, A Voyage to Laputa . . . , in *Gulliver's Travels*, first published 1726.

Introduction

Jonathan Swift's classic fantasy *Gulliver's Travels* satirised the division between those who work with their minds and those who work with their hands: the "scientists" or "philosophers" as against the "engineers" or "mathematical practitioners". Swift was born in Dublin, educated at Trinity College, and became Dean of St Patrick's Cathedral in 1713: his comments, through Gulliver, on the people of Laputa, emphasise the traditional divisions and lack of respect by theorists for users. It is, however, this very "vulgar and mechanick" aspect of Irish instrument making, embodied in the instruments themselves, which was part of the inspiration for this project.

The literary inspiration for this work was D.J. Bryden's monograph on the Scottish scientific instrument trade between 1600 and 1900[1]. Although published as long ago as 1972, it remains a thought-provoking overview of the activity in a geographically remote and, in some ways, culturally retarded locality. The essay has not been superseded: instead, it provided the basis for an in-depth study of a number of individual Scottish businesses[2]. Bryden commented that among "other aspects awaiting further research are the exact relationship between the Scottish trade and the large and important centre of instrument manufacture in London, and a detailed comparison with English provincial centres of a similar size"[3]. However, before this can be done, some sort of description of the trade in all the important localities in these islands should be outlined[4]. We hope that we have provided such an outline for the island of Ireland, which for much of the period under discussion included Britain's second largest city, Dublin, and its own particular markets. We are also aware that this exercise has only produced a preliminary sketch of the Irish instrument trade, and we hope that further study will be undertaken to fill out more of the detail which remains to be uncovered[5].

Parameters

We have had to provide some limits for the scope of the project, in time, geography and in subject matter. The date 1921 seemed to offer itself as a reasonable point to end, as by then the most important instrument maker, Sir Howard Grubb, had had his works removed to England. The nature of instrument manufacture had undergone great changes generally and production had virtually ceased except in specific localities, none of them in Ireland. Thus the birth of the Irish Free State provided an appropriate place for this project to terminate. The twentieth century has its own historiographic problems, and it should be assessed elsewhere by scientific instrument historians other than ourselves. Geographically, we have confined ourselves to the island of Ireland, although we have indicated that makers and retailers were by no means tied to their localities. The first instruments were imported; then skilled workmen migrated to Ireland; subsequently skilled Irish instrument makers emigrated themselves to various destinations. Within Ireland itself, some makers were peripatetic. However, we have looked more carefully at the home markets, and this led us to examine in some detail the major centres of population: Dublin, Belfast and Cork.

This brings us to the tricky definition of "scientific instruments"[6]. We have included in the instrument making category those described in the street directories as "mathematical", "nautical", "optical" or "philosophical" instrument makers; those who appear as "weatherglass-makers" or "barometer makers"; and those with other specific descriptions, such as "electrifying machine maker". We have excluded some groups of specialised instruments, believing their history to be separate to that which we have attempted to outline. Sundial makers have, by and large, been considered outside the scope of this study, along with watch, clock and chronometer makers. This is because the subject has a large and specialised literature of its own, although Irish practitioners have yet to receive treatment beyond a listing of names and dates[7]. On the occasions where these items have been made or retailed by an instrument maker, especially if dated, they have been included. Stone sundials and their makers have been excluded, as the tradition appears to have had more in common with architects and stonemasons[8].

Similarly, medical instruments have been excluded; the cutlers and metal workers who produced them seem to have had even less to do with instrument making than the clock makers. The same can be said for specialist balance makers, scale manufacturers, and makers of weights and measures; these areas both have specialist literatures, but again, as with clockmaking, there is no exclusively Irish work which goes beyond a listing[9].

A more difficult area is that of barometer making. As discussed above, where a maker described himself as "weatherglass maker" we have included him. However, the large proportion of these instruments was produced for use in the home, and properly speaking should be regarded as furniture rather than scientific hardware. We found that we encountered large numbers of barometers, some made by *bona fide* instrument "makers", like members of the Yeates or Mason families; others were produced by specialist glassblowers, who described themselves in street directories as "carvers and gilders and looking-glass makers". Where we have found instruments, we have included the retailer[10].

Another complicated area is the meaning of the word "optician". Initially it meant that the instrument maker was making optical goods, but by the end of the nineteenth century it had acquired its present day connotation and was being used to describe ophthalmic opticians, or "spectacle makers" – who by this date would have been prescribing the spectacles rather than manufacturing them. We have tried to exclude ophthalmic opticians *per se*, but to include those who were also retailers of scientific instruments, and this has meant the inclusion of watch and clockmakers, jewellers, and, in one rather bizarre association, a spermaceti wax manufacturer. By and large, we have included names and dates where there is no real evidence one way or the other.

On the other hand, we have found instruments which have not fitted into the pattern revealed through printed and manuscript sources. These are discussed in the text, but their makers may not appear in the list of makers and retailers, because a single surviving instrument does not necessarily indicate a commercial workshop. An example of this is the Kinsale magnet,

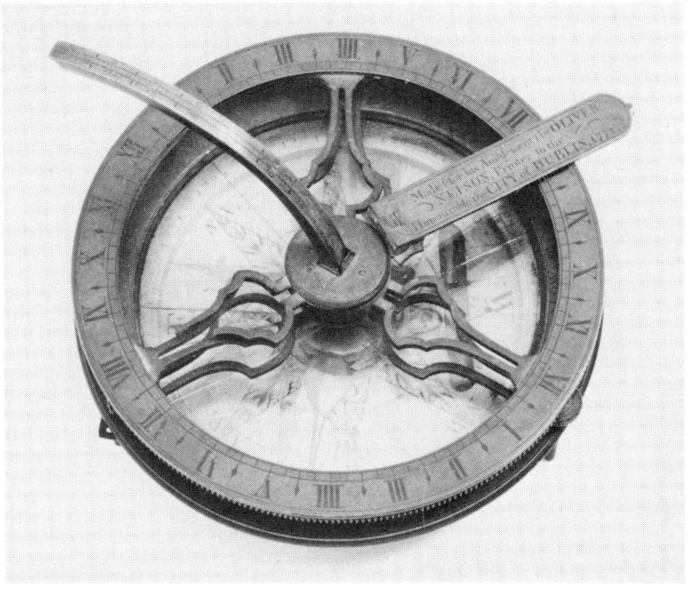

1.1. Universal mechanical equinoctial dial, signed "Made (for his Amuſement) by OLIVER NELSON, Printer to the Honourable the CITY of DUBLIN, 1771". Now in the National Museum of Ireland. Courtesy of Sotheby's.

1.2. Universal mechanical equinoctial dial, signed "Edw. Hunter Fecit"; the case contains a trade card for "Jas Heron Clock & Watch Maker in Newton". Courtesy of Sotheby's.

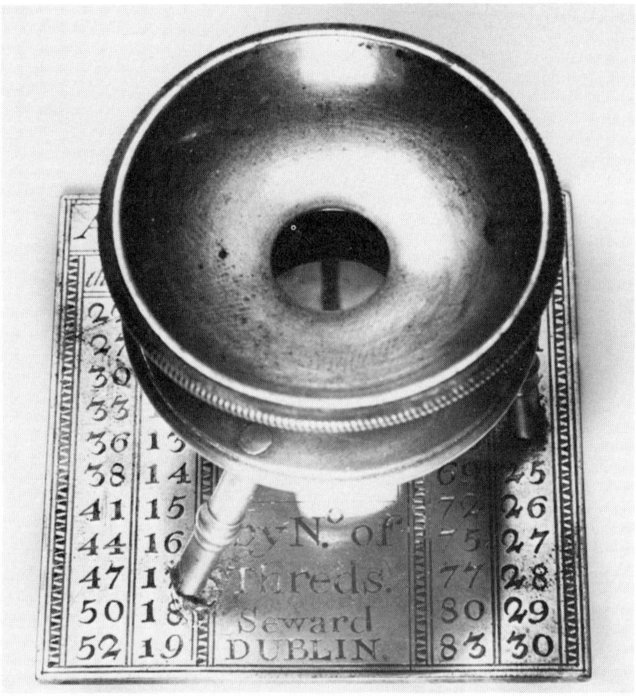

1.3. Linen prover, signed "Seward, DUBLIN". National Museum of Ireland.

made by an amateur, William Lewis, in 1843 (see *Fig.* 5.4 page 78)[11]. Another is the mechanical equinoctial dial signed and inscribed "Made (for his Amuſement) by OLIVER NELSON, Printer to the Honourable the CITY of DUBLIN, 1771" (*Fig.* 1.1)[12]. There is a group of these dials, which all resemble one another, but which appear to have come from a variety of makers: the so far unidentified "Edn. Hunter", who claimed to have made the example which had a trade label for "Jas Heron Clock & Watch Maker in Newtown" (*Fig.* 1.2)[13]; an incomplete, unsigned example[14]; another signed "Made by S. Yeates, 29 Capel St. Dublin"[15]; and two examples by "Seward, Dublin". Seward is known to have signed other instruments, yet his Christian name, address and dates remain untraced[16] (see *Fig.* 1.3). This group illustrates in miniature some of the problems of earlier Irish instruments: who made them, who sold them, and who bought them? It raises some important, if unanswerable, questions: for instance, at what stage in instrument production was the "maker's" name engraved on a piece? How far were clockmakers involved in the instrument trade? How much finishing was done by outworkers? For example, between 1791 and 1796 Robert Burnside of 5 Cope Street advertised in the Dublin street directory as a "shagreen-case maker" and, although probably most of these were intended as jewellery cases, some of his custom surely went to established instrument makers.

Sources

We have used mainly printed sources, although where possible manuscripts have been consulted; no business archive of any Irish instrument making firm was discovered. However, only two such archives survive for Scotland[17], and only a handful in England, so that this seeming dearth was not a great surprise. More difficult was the general lack of manuscript collections, due in part to political disturbances earlier this century[18]. However, the National Library of Ireland's manuscript and printed book catalogues provided an excellent starting point[19]; and we were able to examine manuscript collections relevant to instruments in Trinity College, Dublin, the Royal Dublin Society, the Royal Irish Academy and University College, Dublin.

The principal published source for the lists of instrument makers has been the street directories. We discuss those available for Dublin at the beginning of Chapter Seven, and list those consulted for the remainder of the island in Chapter Eight. The purpose of the directory was commercial, and because Dublin was a large linen trading centre it became the first place outside London in these islands to produce one, in 1751. However, little is known for certain about the methods used to compile street directories. J.E. Norton mentions the editor's personal knowledge, his use of directory agents, the sending of circulars which would lead to inadequate returns, and the use of lists made for other purposes; she comments that the general mobility of the population made directories out of date almost as soon as they were issued[20]. However, it is a reasonably accessible source, easier to work from than manuscripts of census or tax returns: we will turn to directories again in order to put the Irish trade into its context alongside the trade in England and Scotland.

The greatest untapped printed source for the eighteenth and nineteenth centuries must surely be newspapers. The two largest collections of these are held in the National Library of Ireland, and at the British Newspaper Library in London. Newspapers can provide genealogical information, advertisements and announcements, and often add unexpected colour to a business which had merely been a name in a street directory. Our coverage of newspaper archives has been neither consistent nor exhaustive; with few indices, we have relied on serendipidity, and were surprised by the excellent quality of what we found[21]. It remains for some future researcher, ideally based in Dublin, to pursue this source to its natural conclusion.

We have looked at the surviving instruments themselves, whenever possible. These are found in greatest profusion, as one would expect, in Ireland, and we were received with great kindness by owners and guardians of public and private collections[22]. We also examined Irish instruments which have found their way into collections in the United Kingdom: although this group was not numerically large, the diversity was notable. Other traces of instruments were pursued, especially those recorded in auction catalogues. As interest in the history of science and in scientific instruments has grown over the last twenty years or so, it is gratifying to see how the descriptions of instruments have improved in sale catalogues, almost allowing individual pieces to be identified. Looking at the signature and the address together with details of style can help date an instrument; conversely, a dated piece can help to extend a maker's known working life.

Context

The instrument trade in Ireland should be assessed within a variety of contexts: social, economic, scientific. It is probably too early to be able to do more than indicate any of these, and certainly premature to attempt any sort of synthesis. The known people who participated in the trade – and they may have had large or small workforces – still makes a statistically insignificant number in the adult working population, when considered against other merchants dealing in, for instance, linen, or butter. In the eighteenth century, they are better considered with other craftsmen, such as furniture makers (whose individual biographies can be gleaned) or printers. They would have been numerate, literate, and therefore must have been educated or at least apprenticed. Where the information exists, as it does with the Masons and the Yeates, it appears that there was sometimes family tradition involved as well. By the nineteenth century, they were perhaps closer to the communities of engineers and scientists to be found in industry and the universities, but this period of growth was not sustained, and by the end of the century, Irish science was on the decline, the instrument trade with it[23].

One author has attempted to give an idea of the growth of the instrument trade through-

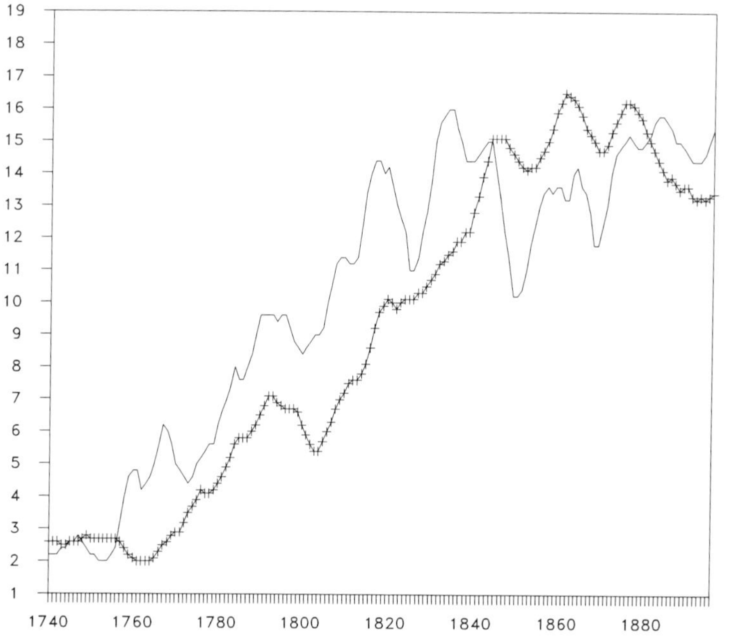

1.4. Graph indicating numbers of known instrument makers in Dublin (solid line) against those in Edinburgh and Glasgow (broken line), derived largely from street directories. Scottish figures have been obtained from D.J. Bryden, Scottish scientific instrument-makers 1600-1900 *(Edinburgh, 1972). Short-term fluctuations have been smoothed by averaging figures over five years.*

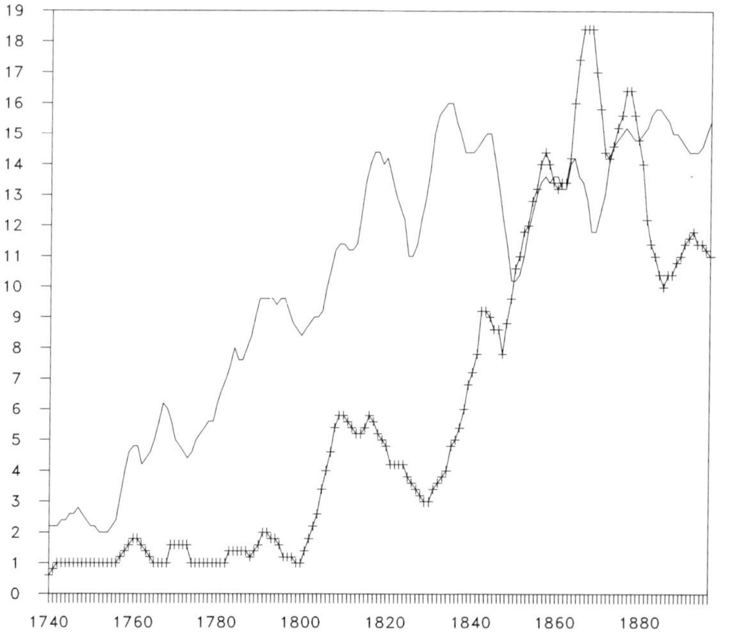

1.5. Graph indicating numbers of known instrument makers in Dublin (solid line) against those elsewhere in Ireland (broken line), derived largely from street directories. Short-term fluctuations have been smoothed by averaging figures over five years.

out these islands for 1700-1840, and of Scotland for 1700-1900[24]. Hackmann admits that "two fundamental problems of this analysis are that we have not been able to distinguish between makers and retailers, and that these figures give no indication of the size of the workforce involved in this trade"[25]. He then goes on to look at census material from 1831 to 1891. Despite the problems with street directories, we should like to compare data gleaned from this source with those used by Bryden (*Fig.* 1.4) and, in another table, show how Dublin makers compared numerically with those in other localities in Ireland (*Fig.* 1.5).

Conclusion

We hope this contribution to the growing literature of the history of scientific instruments will be useful to collectors, curators and all others interested in the surviving artifacts of science. It is an attempt to show that the objects themselves, though mute, can reveal information about the structure of a trade and skills which have all but disappeared. London was perhaps the centre for such activities in these islands but the regions needed instrument makers too.

This work began as a list of names, dates and addresses, then expanded to include a brief introduction, and now attempts to produce a broad outline of events, with thumbnail sketches of the main businesses. We have tried to indicate some European context, in an effort to prove that Ireland was perhaps not quite so far out on the edge of things as might be suggested by glancing at a map. We are, however, acutely aware that this should be regarded as an interim report, because neither the questions we have raised nor the sources examined are exhausted.

References and Notes for Chapter One

1. D.J. Bryden, *Scottish scientific instrument-makers 1600-1900*, Edinburgh, 1972.
2. T.N. Clarke, A.D. Morrison-Low and A.D.C. Simpson, *Brass & Glass: scientific instrument making workshops in Scotland as illustrated by instruments from the Frank collection at the Royal Museum of Scotland*, Edinburgh, 1989.
3. Bryden, *op. cit.*, 1 (ref. 1).
4. A brief survey of projects amassing data on scientific instrument makers in various localities, including those in Europe and in the United States, is given by Willem D. Hackmann, Instrumentation in the theory and practice of science: scientific instruments as evidence and as an aid to discovery, *Annali dell'Istituto e Museo di Storia della Scienza di Firenze*, **10**, 1985, 87-115.
5. For instance, Dr Charles Mollan's work on the Irish National Inventory is producing the locations of many more instruments with an Irish provenance.
6. For a recent discussion, see A.J. Turner, *Early scientific instruments 1400-1800*, London, 1987, 171-230.
7. Geraldine Fennell, *A list of Irish watch and clock makers*, Dublin, 1963. The material gathered by W.A. Seaby, formerly Director of the Ulster Museum, was incorporated in B. Loomes' 1976 supplement to G.H. Baillie's *Watchmakers and Clockmakers of the World*.
8. A nineteenth century dial maker, who came from Ulster, worked mainly in Scotland, and returned to Ireland, was Richard Melville or Melvin: his life and work, as far as can be uncovered, are discussed in Clarke *et al.*, *op. cit.*, (ref. 2). An earlier peripatetic dial maker, also active on both sides of the sea, was John Bonar of Ayr, who is discussed in R.R.J. Rohr, A sun, moon and tidal dial, *Antiquarian Horology*, **16**, 1986-87, 227-232; and A.R. Somerville, The sundials of John Bonar, schoolmaster of Ayr, *ibid.*, 233-242.
9. For medical instruments and their makers, see Elisabeth Bennion, *Antique medical instruments*, London, 1978; for scales and their makers, see J.T. Graham, *Scales and balances*, 2nd edn., Aylesbury, 1986 and M.A. Crawforth, *Weighing coins: English folding gold balances of the 18th and 19th centuries*, London, 1979.
10. A fairly comprehensive list of barometer makers and retailers, as well as an excellent discussion of the subject for the whole of these islands, is Nicholas Goodison, *English barometers 1680-1860*, 2nd edn., Woodbridge, 1977.
11. Now in the Whipple Museum of the History of Science, Cambridge, inventory no. 970.
12. Offered for sale by Sotheby's 10 March 1987 Lot 102, and now in the collection of the National Museum of Ireland.
13. Offered for sale by Sotheby's 25 February 1986 Lot 108. Fennell, *op. cit.*, (ref. 7) gives a James Hunter, watchmaker, Belfast, 1789; Alexander Heron, Newtownards, 1824; William Heron, Donegal, c.1780; William Heron, Newtownards,

1784.
14. Offered for sale by Sotheby's 25 February 1986 Lot 101.
15. Sold by Nouveau Drouot, Paris, The Linton Collection, 9 and 10 October 1980 Lot 123.
16. One example is in the National Museum of Ireland, inventory no. 578-1910; the other was offered for sale by Sotheby's 13 November 1961 Lot 78.
17. They are that of Berry & Mackay of Aberdeen; and James White of Glasgow. Both these firms are discussed in Clarke *et al.*, *op. cit.*, (ref. 2).
18. In particular see Margaret Griffith, Destruction of Public Record Office in 1922, in *A short guide to the Public Record Office of Ireland*, Dublin, 1952.
19. R.J. Hayes *et al.* (eds.), *Manuscript sources for Irish civilisation*, 11 vols., Boston, 1965; *idem.*, *Printed sources for Irish civilisation. Articles in Irish periodicals*, 9 vols., Boston, 1970.
20. J.E. Norton, *Guide to the national and provincial directories of England and Wales*, London, 1950, 45.
21. Rosemary ffolliott, Newspapers as a genealogical source, in D.F. Begley (ed.), *Irish genealogy: a record finder*, Dublin, 1981, 117-138, gives general information about locations and availability of Irish newspapers and indices. Other than those mentioned by ffolliott, we have consulted Henry Farrar, *Irish marriages: being an index to the marriages in Walker's Hibernian Magazine 1771 to 1812*, London, 1897; [Anon.], *Index to births, marriages and deaths in . . . the Hibernian Chronicle October 1769-April 1802*, vol. I, 1769-1775 (typescript in the British Newspaper Library); [Anon.], *Index to birth, marriage and death entries in the Belfast Newsletter 1730-1800* (typescript in the Linenhall Library); and the computer index, still being compiled by Dr John Green, of the *Belfast News-Letter* 1753-1800.
22. Dr Charles Mollan is compiling a National Inventory of instruments in Ireland, which will add greatly to knowledge about the trade. An "Interim Report", listing 1750 instruments, is available from him at the Royal Dublin Society, Dublin 4. We would also like to thank Dr Willem Hackmann for information about Irish instruments contributed to the forthcoming Inventory for the British Isles, to be published jointly by the Royal Society and the Science Museum, London, and to Dr R.G.W. Anderson for making this pre-publication information available to us.
23. Gordon L. Herries Davies, Irish thought in science, in Richard Kearney (ed.), *The Irish mind: exploring intellectual traditions*, Dublin, 1985, 294-310.
24. W.D. Hackmann, The nineteenth-century trade in natural philosophy instruments in Britain, in P.R. de Clercq (ed.), *Nineteenth-century scientific instruments and their makers*, Leiden, 1985, 74-76. Hackmann's figure 1 uses information "mainly based" on instrument makers listed in E.G.R. Taylor, *The mathematical practitioners of Hanoverian England 1714-1840*, Cambridge, 1966, and Bryden, *op. cit.*, (ref. 1); and Hackmann notes that "these figures have to be used with extreme caution".
25. *Ibid.*, 76.

CHAPTER TWO

SCIENTIFIC INSTRUMENTS IN IRELAND TO THE END OF THE SEVENTEENTH CENTURY

J.E. Burnett

Introduction

Prehistoric man needed some knowledge of the natural world in order to survive. How deep that knowledge was, and how extensive, we can never know. Indeed, it is possible to notice only a few glimpses of a systematic interest in nature in Ireland before the seventeenth century AD.

Man was certainly present in Ireland from the seventh millenium before the birth of Christ. Ireland's most spectacular archaeological remains are the passage-graves of the Boyne Valley, which were built in the fourth millenium BC. At the most famous of them, Newgrange, carbon-14 determinations indicate a date of 3300 BC[1]. Externally, Newgrange is an earthen mound; a long, narrow passage leads into the heart of it, where there is a small chamber. At the winter solstice the sun shone directly into the chamber[2]. Clearly, the people who built this tomb had some knowledge of astronomy. An analysis of thirty stone circles in Cork and Kerry showed that seventeen contained orientations on the rising or setting of the sun, moon and Venus[3]: those circles which contain astronomical alignments demonstrate an understanding of a particular aspect of nature, and it is not unreasonable to regard them as giant scientific instruments, instruments laid out by man to enable men to know more easily when certain days of the year occurred.

British-born St Patrick arrived in Ireland as a missionary in 432 AD, or thereabouts. He spread the Christianity which had arrived there before him, and at Tara – ten miles south west of Newgrange – confronted the high-king, Laoghaire, and silenced the druids. Through St Patrick's spiritual conquest, Ireland gained access to the scientific learning of classical antiquity. St Patrick is linked to our present account in another way, for he chose Armagh as the site of his episcopal see. In the eighteenth century an ecclesiastically-funded astronomical observatory was established there.

The Romans had left England before St Patrick began his mission in Ireland. The Holy City was sacked, the Empire disintegrated, and continental Europe entered upon an age of barbarism in which the conspicuous figures are roughnecks such as Theodoric the Ostrogoth and Clovis the Frank. From about 600 AD the disruption and uncertainty was penetrated by Irish missionaries. Initially they founded monasteries and spread the faith, but after 800 AD Irish scholars who had travelled to the Continent became more prominent. Through the efficient lines of communication maintained by the Church, Ireland remained abreast of medieval knowledge and scholarship[4]. Astronomy, in particular, must have been a part of that learning. Evidence of this is supplied by an astronomical rotula at the beginning of a Gaelic astronomical text (*Fig.* 2.1)[5]. The text itself was written by Messahalah, an Alexandrian Jew, in the late eighth century; it was translated from Arabic to Latin in the thirteenth century[6], and the manuscript in question dates from the fourteenth or fifteenth century[7].

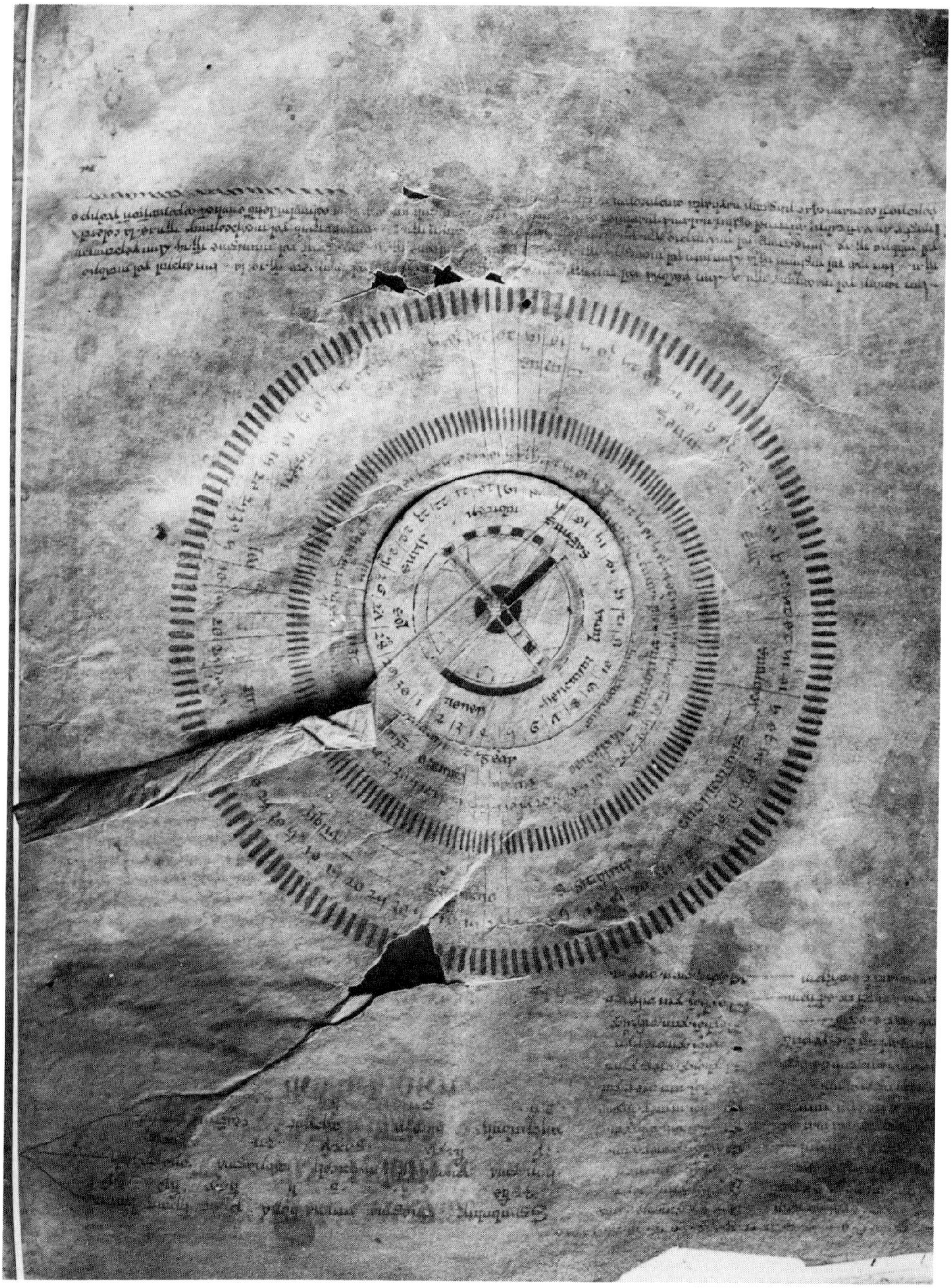

2.1. Gaelic astronomical text, with rotula, possibly fourteenth Century. Royal Irish Academy.

Most of the present book is concerned with instruments which were created in response to the demands of traditions of natural knowledge and technological manipulation of nature which began to change rapidly about 1600, and have continued to change ever since. It must be emphasised that the text of Messahalah described above was written in Gaelic, and that Gaelic scholarship continued with only a distant relationship to "mainstream" intellectual activity in England.

In classical antiquity instruments for surveying and astronomy were developed and knowledge of their use passed to the muslim civilisation of the Middle East. We may regard the European tradition of instrument making as having begun in Moorish Spain. The earliest known medieval European instrument is a planispheric astrolabe constructed by an Arab craftsman in 1026 or 1027[8]. From the eleventh century a continuous history of precision engraving on brass can be perceived. The most important of early instruments was undoubtedly the astrolabe. Although there are several astrolabes in collections in Ireland[9], there is none which has an Irish provenance, in the way, for instance, that the French instrument used by the Scottish cartographer Robert Gordon of Straloch (1589-1661) has a Scottish provenance[10]. A mariner's astrolabe was found on Valencia Island in 1845, and it has been assumed that the instrument came from one of the wrecked ships from the Spanish Armada trying to reach Spain in 1588[11].

Ireland's geographical position on the edge of the known Western world has made her a prey to her more aggressive land-hungry neighbours. As Beckett has said:

> An island people without command of the sea was easily open to attack, and Ireland's broad, slow-moving rivers offered convenient access from the coast to the interior . . . Though the interior can be reached easily from the coast, Ireland has no natural focal point, no great crossing-place of routes, no centre from which influence spreads naturally to the circumference. This was one of the facts which helped to prevent the establishment of political unity from within. Dublin is essentially an invaders' capital; founded originally by the Norsemen it became the centre of Anglo-Norman power in Ireland chiefly because of its position on the east coast, within easy reach of the ports of Bristol and Chester[12].

Dublin has been the capital of Ireland – or of as much of Ireland as was under a central control – for nine centuries. But the most significant factor in Ireland's geography is the relative position of Great Britain, acting as a barrier between her and European influences, and using her as a colonial testing ground for imperial designs. England, for her part, long held the fear that Continental powers could seize Ireland and use her as the "back-door" for an invasion. This uneasy relationship has coloured much of Ireland's history, and because it affected her economic and social history, influences the history of instrumentation in Ireland.

Activity in natural philosophy as opposed to "mathematical practice" in Ireland, like that in neighbouring Scotland, lagged appreciably behind the leading countries of Europe[13]. In part, this can be accounted for by recurring rebellion and settlement – the series of "plantations" by colonists from England and Scotland who settled on confiscated Irish land from Norman times to the Glorious Revolution. Military and political upheaval tends to mitigate against most cultural preoccupations. It was in a lull in this stormy political climate, between the Munster rising of 1579-83[14] (which was aided by the Spanish, whom Queen Elizabeth was fighting in a number of theatres of war as far apart as the Netherlands and the Caribbean[15]) and the more threatening Ulster Rebellion of 1595-1603[16], that Ireland's first university was founded in 1592 as Trinity College, Dublin[17]; and, although its initial purpose was to provide a supply of well-trained clergy for the established church, it did form some focus for limited scientific activity over the next century or so: some medical skills were taught, and in 1612 a series of mathematical lectures given. From its foundation, a surprisingly large proportion of scientific works was bought for the library[18].

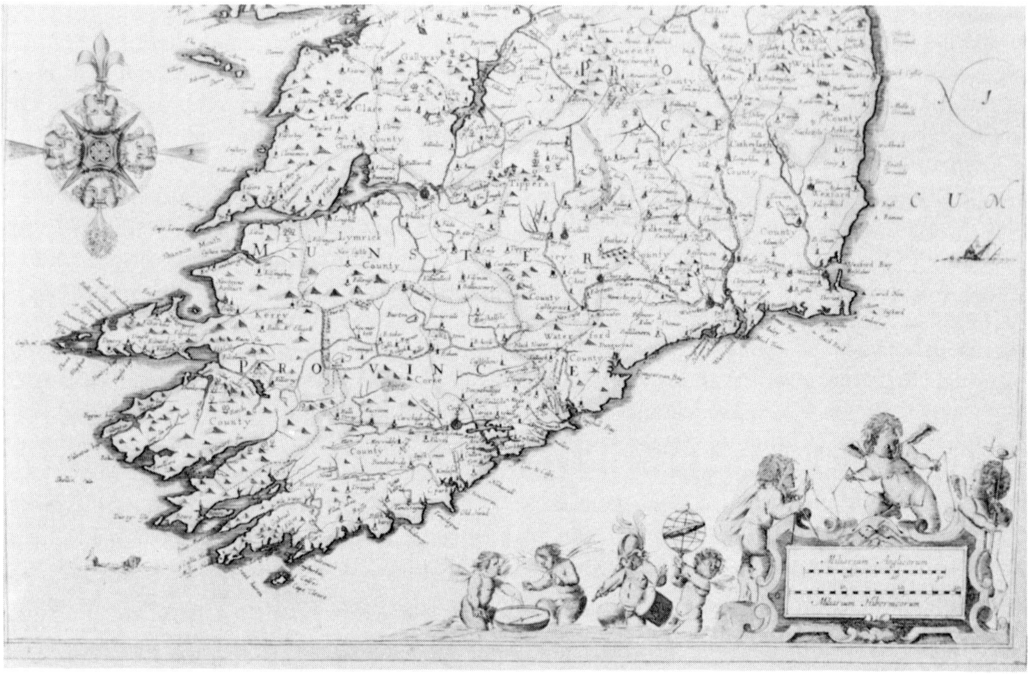

2.2. Cartouche from William Petty's map of Ireland, 1685. Royal Dublin Society.

Early instruments

It appears, however, that the earliest instruments used in Ireland were not made locally. A possible rare exception to this was the instance of Miles Symner, a major in the parliamentary army, who in 1652 became professor of mathematics at Trinity College; earlier, probably in the 1640s, he had made for himself a six foot quadrant[19]. He determined the latitude of Dublin to be 53° 20′[20], the figure found on most eighteenth century Dublin-made sundials.

The first activity in Ireland which required the use of instruments appears to have been surveying, an activity which began there in Tudor times, if not earlier[21]. The first survey made using instruments (as opposed to the first map[22]) of a significant proportion of Ireland, was that planned and superintended by Sir William Petty in 1655 and 1656[23], and popularly referred to as the "Down Survey" since everything was written down. Petty, later a founding member of the Royal Society of London, had gone to Ireland as Physician to the Forces, having previously held the post of Professor of Anatomy at Oxford[24]. His survey was carried out at the instigation of the English Cromwellian government, to map and measure the large but discontinuous portion of Ireland forfeited by landowners sympathetic to the 1641 rebellion, which was put down with great brutality by Cromwell's army in 1649[25]. As one of the political consequences of the rebellion, which had left England's "back door" open to the threat of foreign invasion for some seven and a half years – fortunately, the rest of Europe was too busily embroiled in the Thirty Years' War to pay much attention – the survey needed an army of surveyors, traversing with chain and compass the boundaries of townlands, parishes and baronies in the provinces of Ulster, Leinster and Munster[26].

John Aubrey summed up Petty's surveying: "The Kingdome of Ireland he hath surveyed, and that with that exactnesse . . . and those that he employed for the Geometricall part were ordinary fellowes, some (perhaps) foot-soldiers, that circumambulated with their *box and needles*, not knowing what they did, which Sir William knew right well how to make use of"[27].

The chief instruments used were circumferentors. In Petty's own words:

> the method and order used by the said Petty in this vast worke was as followeth, viz: Whereas surveyors of land are commonly persons of gentile and liberall education, and theire practise esteemed a mistery and intricate matter, farr exceedinge the most parte of mechanicall trades, and withall, the makeinge of theire instruments is a matter of much art and nicety, if performed with that truth and beauty as is usuall and requisite. The said Petty, consideringe the vastness of the worke, thought of dividinge both the art of makeinge instruments, as alsoe that of usinge them into many partes, vizt., one man made onely measuringe chaines, vizt., a wire maker; another magneticall needles, with theire pins, vizt., a watchmaker; another turned the boxes out of wood, and the heads of the stands on which the instrument playes, vizt., a turnor; another, the stands or leggs, a pipe maker; another all the brasse worke, vizt., a founder; another workman, of a more versatile head and hand, touches the needles, adjusts the sights and cards, and adaptates every peece to each other.
>
> In the meane tyme scales, protractors, and compasse-cards, beinge matters of accurate division, are prepared by the ablest artists of London[28].

This is one of the earliest descriptions of the division of labour within the instrument trade, showing that there were various specialisms within it, and also that some tasks were so skilled, particularly dividing, that the pieces had to be imported. Unfortunately, Petty does not name his London suppliers, but some of the squared-paper original plottings survived until the mid-nineteenth century; they were graduated at the edges for protraction, printed from engraved copperplates, and in the margin bore the date 1654 and the name H. Sutton[29]. Henry Sutton was a renowned London instrument maker, particularly noted for the high quality of his engraving[30].

*2.3. The earliest known surviving Irish scientific instrument: a wooden circumferentor and sundial, signed "W.R. Dublin. fe. Latitude 53° 20′ *1667*". Museum of the History of Science, Oxford.*

The earliest surviving Irish-made instruments are surveying instruments. The earliest of all is a wooden circumferentor, or surveying compass, now in the Museum of the History of Science, Oxford (*Fig.* 2.3)[31]. It is signed "W.R. Dublin. fe. latitude 53° 20.' *1667*" and so dates some ten years after the Down Survey.

The compass card of this instrument is printed from an engraved plate, and it incorporates a sundial. The engraving is of a high quality, and is very similar to that of contemporary London engravers. For example, Walter Hayes of London had John Twysden write a booklet on the planisphere which Hayes made, and Hayes himself supplied the engravings. There is a similarity between Hayes' style and that of the Irish circumferentor, and this is confirmed by an instrument made by Hayes which bears some resemblance to it[32]. It would be easy to believe that Hayes – or rather his workmen – had done the engraving for the W.R. instrument. The problem of establishing which individual actually carried out a particular piece of engraving is indicated clearly by Twysden. Twysden referred to "my good Friend, Mr Walter Hays, who doth employ and direct the best, or as good workmen as any one, to perform what his Age or Infirmity makes him unable to perform with his own Hands. In this he is to be commended, that you will very seldom find him out of Doors, and to be sought in Ale-Houses – or Taverns, where many a good Workman spends much of his time"[33].

The work of Henry Sutton is similar to the engraving of the W.R. instrument, and paper instruments made by Sutton survive[34]. The compass card on the W.R. circumferentor is almost identical to the one illustrated on Sutton's trade card of 1654 (*Figs* 2.4a & b)[35]. The layout of the two, with a compass rose in the centre, followed by the signature, a calendar scale, part of a stereographic projection, and 360 degree scale, are identical. The outermost scale on the Sutton trade card is of degrees, separated into four quadrants, each running 0-90°. The W.R. instrument has been mutilated and bears a manuscript scale which runs from zero at north to 120 at north, i.e. it is divided into units of three degrees: the purpose of this scale has not yet been explained. Small differences can also be detected in the engraving of the two cards: in the fleur-de-lys and the suns, and in the use of dotted lines for two of the lines of declination on the Sutton card. The Sutton card has additional lines indicating sunrise and sunset. It can be asserted with confidence that the two cards were not printed from the same plate. Perhaps the safest conclusion is the rather weak one, that this type of dial was being made by a number of craftsmen in London in the 1660s, and that one of them engraved the plate for W.R.

But who was W.R.? There is no record of a watch or clockmaker or printer with these initials in seventeenth century Dublin. It is possible that he was a London-trained mathematical instrument maker who came to try his luck in Ireland in the comparative prosperity which followed the Restoration of the Stuart monarchy – but the market cannot have been large. Did W.R. have some other skill or profession? If so, the instrument could have been made at his request in London, and sold in Dublin. With this in mind, the name of Sir William Robinson (c. 1643-1712) must be mentioned. Robinson is remembered as the architect of the Royal Hospital at Kilmainham, on the outskirts of Dublin. He was an Englishman, who is first known of in Ireland in 1674, though he may have been there in 1670[36]. The nature of his education, or his whereabouts in the 1660s, are unknown. In 1674 he was described by the Earl of Essex, the Lord Lieutenant of Ireland, as being "a very ingenious man & well skilled in some parts of ye Mathematicks"[37]. Robinson's activities in Ireland were mostly military: he was Surveyor-General from 1671-1700[38].

A second seventeenth century brass instrument, signed "Johannes Lewis Dublini fecit anno Domini 1688", survives at Maynooth College (*Fig.* 2.5). One John Lewis was a watchmaker, a member of the Goldsmiths' Company of Dublin in 1679. He died in 1680[39], so that the maker of this instrument is probably another man with the same name. Two other instruments, one dated 1684, have been recorded with his signature[40].

It is perhaps significant that the two earliest instruments are surveying instruments. The

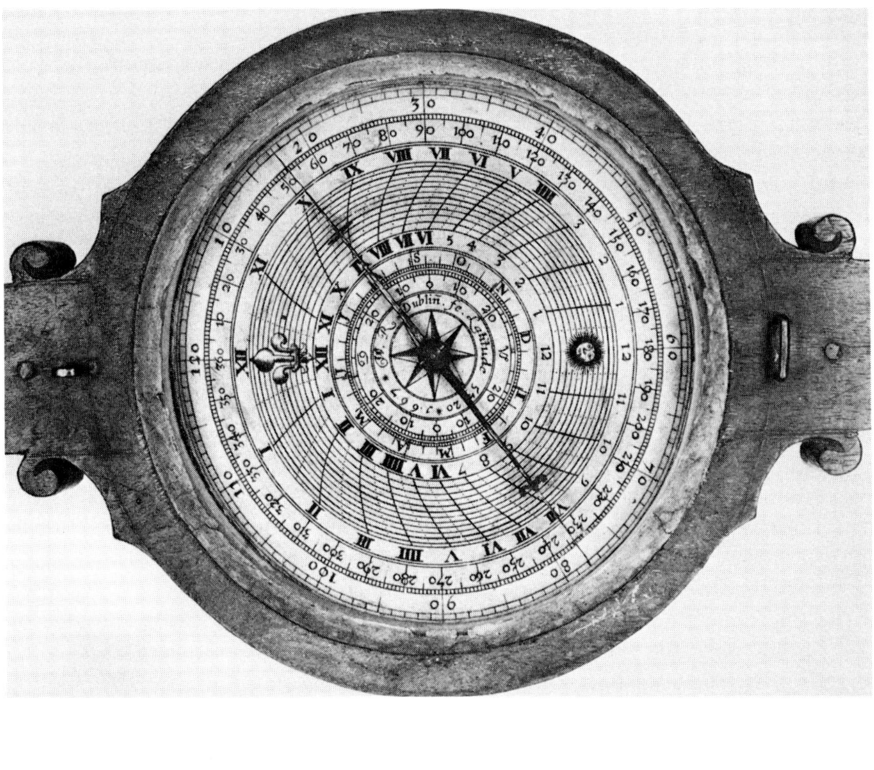

2.4.(a) *Detail of wooden circumferentor, signed "W.R. Dublin. Fe Latitude 53°20' * 1667 *". Museum of the History of Science, Oxford.*

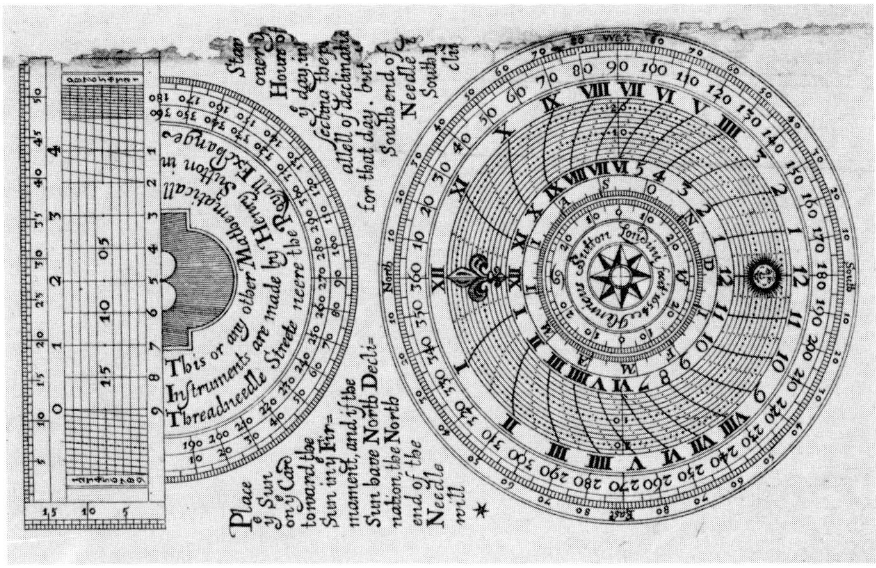

2.4.(b) *Trade card with compass and sundial signed "Henricus Sutton Londini fecit 1654". Trustees of the Science Museum, London.*

*2.5. **Brass circumferentor, signed "Johannes Lewis Dublini fecit anno Domini 1688".** St Patrick's College, Maynooth.*

circumferentor was to become a far more common instrument in Ireland than in England[41]. A parallel can be drawn, perhaps, with that of early American instruments, which were initially imported from Europe, and because there was no ready local supply of brass the earliest American-made instruments were constructed in wood[42]. Circumferentors were also particularly common in America; initially wooden, latterly brass[43]. In the case of Dublin, few instruments were made of wood, as the growth of the city and its prosperity, besides its proximity to Britain, meant that glassworkers, brass founders and, in general, skilled craftsmen – including glass blowers from the Continent – came to settle there from the mid eighteenth century.

The Dublin Philosophical Society, founded in another period of political stability in 1683, was not based on any long tradition of scientific activity, and Hoppen states categorically that "the scientific inspiration for the Dublin Society came almost entirely from England"[44]: there was not the sort of home-grown stimulus to provide any mathematical practitioners with the business needed to set up a local trade. The leading figure of the Dublin Philosophical Society was William Molyneux, whose correspondence with the English Astronomer Royal John Flamsteed reveals the telling comment in September 1681: "living here in a kingdom barren of all things, but especially of the ingenious artificers, I am wholly destitute of instruments I can rely on"[45]. That year Molyneux placed orders for instruments through Flamsteed, with London makers; in 1685 he commissioned Richard Whitehead of London to construct a telescopic dial which Molyneux had himself designed[46]. The Dublin Society itself did not follow its founder's advice which was that experiments were to form the basis of its business; in part this was because its meeting places were too limited, so that the members were obliged to discuss ideas rather than witness demonstrations[47]. The Society itself disintegrated when most of its members were forced to flee the country with the accession of the Catholic King James II; however, most returned to Dublin after the Williamite Revolution by 1692. The Society's revival between 1693 and 1697 was brief. Molyneux himself abandoned natural philosophy for politics and, despite a third revival by Molyneux's son Samuel in 1707, the Society then lasted less than a year[48].

References and Notes for Chapter Two

1. Aubrey Burl, *The stone circles of the British Isles*, New Haven and London, 1976, 241.
2. M.J. O'Kelly, *Newgrange: archaeology, art and legend*, London, 1982, 123-126.
3. Burl, *op. cit.*, 222-223 (ref. 1); J. Barber, The orientation of the recumbent stone circles of the SW of Ireland, *Journal of the Kerry Archaeological and Historical Society*, **6**, 1973, 26-39.
4. W.B. Stanford, *Ireland and the classical tradition*, Dublin, 1976, 182-185.
5. Messahalah, *An Irish astronomical tract*, edited by Maura Power, London, 1914, v (Irish Texts Society, vol. 14). The manuscript itself is in the Royal Irish Academy, MS BH11 (Stowe Collection).
6. *Ibid.*, ii.
7. *Ibid.*, xi.
8. R.G.W. Anderson, *The mariner's astrolabe*, Edinburgh, 1972, 34; R.W. Plenderleith, Discovery of an old astrolabe, *Scottish Geographical Magazine*, **76**, 1960, 25. Now in the National Museum of Scotland, inventory no. T1959.61.
9. The Chester Beatty Library and Gallery of Oriental Art has a collection of Turkish astrolabe-quadrants and a single astrolabe; these were brought to Dublin in 1950. There is an Indian astrolabe in a private collection, and one signed by Erasmus Habermel in the National Museum of Ireland.
10. A.S. Cumming, Gordon of Straloch's astrolabe, *Scottish Geographical Magazine*, **42**, 1926, 79-82; A.R. Hutchieson, Bequest to the Royal Scottish Museum: astrolabe of Robert Gordon of Straloch, *Mariner's Mirror*, **34**, 1948, 122-3. Now in the National Museum of Scotland, inventory no. T1947.27.
11. Now in the collection of the National Maritime Museum, Greenwich, inventory no. A55/54-13. Described in Anderson, *op. cit.*, 22-23 (ref. 8).
12. J.C. Beckett, *A short history of Ireland*, 6th edn., London, 1986, 9. Other helpful accounts, particularly of Ireland's economic history, which have been used as general background are: L.M. Cullen, *Life in Ireland*, London, 1979; L.M. Cullen, *The emergence of modern Ireland 1600-1900*, Dublin, 1983 and L.M. Cullen, *An economic history of Ireland since 1660*, 2nd edn., London, 1987.
13. D. J. Bryden, *Scottish scientific instrument-makers 1600-1900*, Edinburgh, 1972, 1.
14. Beckett, *op. cit.*, 50-51 (ref. 12); Edmund Curtis, *A history of Ireland*, 6th edn., London, 1981, 198-199.
15. S.T. Bindoff, *Tudor England*, Harmondsworth, 1971, 277-288.
16. Beckett, *op. cit.*, 55-60 (ref. 12).
17. R.B. McDowell and D.A. Webb, *Trinity College Dublin 1592-1952: an academic history*, Cambridge, 1982.
18. K. Theodore Hoppen, *The common scientist in the seventeenth century: a study of the Dublin Philosophical Society 1683-1708*, London, 1970, 56-61.
19. *Ibid.*, 62; Rolf Loeber, *A biographical dictionary of architects in Ireland 1600-1720*, London, 1981, 103-105.
20. Discussed by J.H. Andrews, Science and Cartography in the Ireland of William and Samuel Molyneux, *Proceedings of the Royal Irish Academy*, **80C**, 1980, 213-250, n. 36.
21. J.H. Andrews, *Plantation acres: an historical study of the Irish land surveyor*, Belfast, 1985, 1-51.
22. As the most westerly part of the known world, Ireland was of interest from ancient times and appeared in Ptolomy's map of the British Isles, first published at Bologna in 1477: for a resumé of Irish mapping history see R.V. Tolley, *Maps and map makers*, 4th edn., New York, 1970, 91-95.
23. Previous surveys of smaller portions of Ireland had been carried out, usually by military engineers with the series of British attempts to colonise them. See Andrews, *op. cit.*, 28-77 (ref. 21).
24. Biographical details of Petty are to be found in Sir Irvine Mason and A.J. Youngson, Sir William Petty F.R.S. (1623-1687), in Sir Harold Hartley (ed.), *The Royal Society: its origins and founders*, London, 1960, 79-90; a recent review is by J.H. Andrews, Sir William Petty: a tercentenary reassessment, *The Map Collector*, 1987, 34-39.
25. Christopher Hill, *God's Englishman: Oliver Cromwell and the English Revolution*, Harmondsworth, 1975, 106-118.
26. Andrews, *op. cit.*, (ref. 20).
27. John Aubrey, *Brief lives*, edited by Oliver Lawson Dick, Harmondsworth, 1972, 400.
28. [William Petty], A brief accompt of the most materiall passages relatinge to the survey managed by Dr Petty in Ireland, anno 1665 and 1656, in Thomas Aiskew Larcom (ed.), *The history of the survey of Ireland commonly called the Down Survey by Dr William Petty AD 1655-6*, Dublin, 1851, xiv.
29. Larcom, *op. cit.*, 324 (ref. 28); Andrews, *op. cit.*, 65 (ref. 21).
30. E.G.R. Taylor, *The mathematical practitioners of Tudor and Stuart England 1485-1714*, Cambridge, 1954, 220.
31. Museum of the History of Science, Oxford, inventory M20.
32. John Twysden, *The use of the general plani-*

sphere called the analemma, London, 1685.

33. In a private collection.
34. For example, a Gunther's quadrant, a magnetic azimuth dial with surveyor's compass, and a diptych dial: J.A. Bennett, *The divided circle: a history of instruments for astronomy, navigation and surveying*, London, 1987, 75 and 80; Harriet Wynter, *A catalogue of scientific instruments*, 2 (1), 1974, 2-3.
35. H.R. Calvert, *Scientific trade cards in the Science Museum collection*, London, 1971, no. 387; illustrated in George H. Gabb, The astrological astrolabe of Queen Elizabeth, *Archaeologia*, **86**, 1936, 101-103, p. 102.
36. Loeber, *op. cit.*, 88-97 (ref. 19).
37. British Library, Stowe MS 214, f.259, quoted by Loeber, *op. cit.*, 92 (ref. 19).
38. Loeber, *op. cit.*, 89-90 (ref. 19).
39. Brian Loomes, *The early clockmakers of Great Britain*, London, 1981, 357.
40. A compass dial, signed "Johannes Lewis Dub" is in a private collection; we are grateful to A.J. Turner for this information. A brass sundial signed "Johans Lewis, Dublini fecit, latt 53d 20m, anno 1684" was offered for sale at Christie's 12 October 1965 Lot 7.
41. Bennett, *op. cit.*, 149 (ref. 34).
42. S.A. Bedini, *Early American scientific instruments and their makers*, Washington D.C., 1964, 65-129.
43. Bennett, *op. cit.*, 149 (ref. 34). Deborah Jean Warner, The surveyor's compass, *Rittenhouse*, **1**, 1987, 64-74.
44. Hoppen, *op. cit.*, 3 (ref. 18).
45. Quoted in *ibid.*, 93, n.91.
46. *Ibid.*, 113-114. This was considered a sufficiently successful instrument for the authorities of Trinity College, Cambridge, to purchase an example in 1703; D.J. Price, the early observatory instruments of Trinity College Cambridge, *Annals of Science*, **8**, 1952, 3.
47. Hoppen, *op. cit.*, 71 (ref. 18).
48. J.G. Simms, *William Molyneux of Dublin*, edited by P.H. Kelly, Dublin, 1982, 44-46.

CHAPTER THREE

SCIENTIFIC INSTRUMENT MAKING IN DUBLIN, 1700-1830

A.D. Morrison-Low and J.E. Burnett

Introduction

The market for instrument making has been discussed by Bryden[1] and Bedini[2], who define three types of customer: the "scientific", the "professional" – which can be subdivided again into "teachers" and "practitioners" – and the "dilettante". All these markets were to be found in Ireland from the time of Petty, but the largest part of the custom from the Down Survey until the mid-nineteenth century would appear to be that of the "professional" customer outside teaching. Judging by the conclusions reached by a recent publication on surveying[3], and the number of surviving instruments from this period, land surveying was the profession which demanded the most from the instrument maker at this time.

Stokes and the early eighteenth century

The first Dublin instrument maker about whom anything substantial is known is Gabriel Stokes (1682-1768)[4]. He was the son of a Dublin tailor whose family came from Gloucestershire, where they owned land. He attended the King's Hospital School in Dublin, and served an apprenticeship starting in 1696 with Joseph Moland, a notable surveyor who was then employed at the same school as steward and teacher of navigation. It would appear that Stokes learned his surveying skills from Moland, but the source of his instrument making knowledge remains a mystery. His map making and instrument making activities seem to have occurred simultaneously: his first signed map is dated 1716[5], and there is evidence that he was repairing instruments the previous year. Other maps and surveying activity by Stokes are as follows: 1723, map of Affane, Co. Waterford. MS[6]; 1725, plan of Dublin Harbour. MS[7]; 1733, surveying in Kerry and Tipperary for Trinity College[8]; surveying in Dublin, also for Trinity College[9]; 1750, map of Dublin County, engraved by S. Wheatley[10]; pre-1755, map of Ireland for the Post Office, known to have been lost by 1755[11]. The chronological spread of these maps suggest that he practised simultaneously as an instrument maker and surveyor, and did not graduate from one to the other.

Stokes received a grant of arms for "skill in his profession", by which was meant mathematical instrument maker, the only description he gave of himself when witnessing a deed in 1720[12]; his crest was a surveyor's forestaff. His shop was at the "Sign of the Dial" in Essex Street from at least 1731 until 1763[13]. He was Deputy Surveyor-General of Ireland from 1748 to 1752, during which time he introduced what Andrews has characterised as "a major administrative innovation"[14]: the certificate of competence awarded to private surveyors. Certainly surveying seems to have been one of his main lines in mathematical instrument making. At least one brass surveyor's compass marked with Stokes' name has been recorded[15].

In Robert Gibson's *Treatise of Practical Surveying*, of which the second edition was pub-

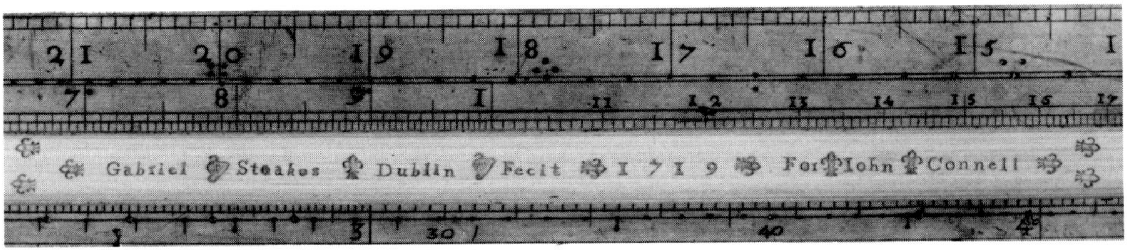

3.1. Detail of wooden Gunter rule, signed "Gabriel Stoakes Dublin Fecit 1719 for Iohn Connell". Whipple Museum of the History of Science, Cambridge.

lished in 1767, Stokes advertised that he sold all the mathematical instruments mentioned in the book – chain, circumferentor, semicircle, plane table and protractor – and "a small instrument . . . much more portable than a Theodolite" designed and made by himself, and called a "Pantometron"[16].

Stokes was the author of *A scheme for effectually supplying every part of the City of Dublin with pipewater without any charge for water engines or any water forcers, by a close adherence to the law of sanitation and the principles, rules and experiments of Hydrostaticks*, published in 1735. His interest in both water supply and instrument making is similar to that of the Scotsman George Sinclar[17], who sold barometers and wrote on hydrostatics, and engineered the water supply of Edinburgh. A 36-inch wooden Gunter sliding rule is stamped: "Gabriel Stoakes Dublin Fecit 1719 For Iohn Connell"[18] and is his earliest dated piece – he would have been about 37 when he made it (*Fig.* 3.1). It is decorated beside the signature with a pattern made from two stamps: one, a fleur-de-lys and the other an Irish harp. It is not, however, the earliest evidence of Stokes' work. In the Muniments of Trinity College, Dublin, is a bill he submitted in 1715. Since this is the earliest known bill of an Irish instrument maker, we give it below in full[19].

Trinity College Dr

	£	s	d
For repairing the divisions of the great Quadrant	2	15	6
for New cutting the rack in the Edge	2	6	0
for repairing the Ingien work of the Index & for Two New handles	1	17	0
for 7 New Screws	0	3	6
for one New Iron Screw to the frame	0	1	0
for a brass Eye piece & upper part to the Tellescope of the Quadrant	0	10	0
for a New Plummett & Silver Wier	0	5	0
for a New Center Pin	0	0	6
for Painting	0	3	0
for Cleaning & Rectifiing	1	4	0
for Carriage Twice	0	1	0
To Mr Rose for four Glasses	1	0	0
	£10	6	6

Aprill 20th 1715
Then Recd. in full the Contents of the Above Bill
Gab. Stokes

The fact that Stokes could re-divide the quadrant, and re-cut its rackwork, indicates that he was a skilled mathematical instrument maker, and raises the apparently unfathomable question – where did he learn these skills? One can only answer limply – London, probably.

The quadrant in question was probably the one purchased by Trinity College at some time between 1685, when they began to equip a small observatory, and 1687[20]. In the latter year Miles Symner's successor as Professor of Mathematics, St George Ashe, wrote[21]:

> We have erected an observatory in the colledge, tho' our apparatus of instruments is but small yet; an accurate crop quadrant of 3 feet radius with all the new invented furniture we have, with which by observing the meridian attitudes of the sun, especially at the late solstice, we have exactly enough setled the latitude of this place . . .

From the above description it would appear that the Trinity quadrant repaired by Stokes is like the quadrant described by Robert Hooke in his *Animadversions on the first part of the Machina Coelestis of . . . Johannes Hevelius* (1674), in that the Dublin instrument is known to have had a "rack in the Edge" and "Ingien work of the Index"[22]. Indeed, it is conceivable that the construction of the quadrant was organised from London by Hooke himself: Hooke is known to have sent a quadrant with telescopic sights to the Irish astronomer Henry Osborne of Drogheda, who has not been noticed after 1656[23] and whose instrument may have been passed to Trinity.

Stokes' bill is endorsed on the reverse:

> I frequently called at Mr. Stoakes's while he was at work on repairing the Quadrant belonging to ye Colledge, and I always observ'd he had been particularly carefull in amending what was amiss; and it appeared to me when the work was finish'd, & the observations made with the instrument afterwards, that he had perfected his work & adjusted the Quadrant with Skill & exactness: and in my opinion the Several Articles of ye work done by hime are moderately rated
> 6th May 1715 Tho. Burgh

Thomas Burgh was not, as one might expect from this endorsement, a Fellow of Trinity. He appears, rather, in the role of technological polymath. Burgh, son of the Bishop of Ardagh, had been an undergraduate at Trinity in the 1680s[24], and served as a military engineer in the Low Countries in the early 1690s. He returned to Ireland, and succeeded Sir William Robinson as Surveyor-General in 1700. As an engineer, he constructed barracks and fortifications, advised Dublin on its water-supply and its harbour, and he had a share in a coal mine. As a public man he was High Sheriff of Co. Kildare and a Member of Parliament. As an architect, he built the library at Trinity College (1712-33), where the Stokes bill now lies, and also the new chemical laboratory and anatomical theatre of 1710-11[25].

Two other bills dating from the first half of the eighteenth century pose further questions. The first is from William Vream to Richard Helsham[26]. Vream has been recorded trading in London between 1710 and 1717: he is known to have made air-pumps to the design of Francis Hawksbee the elder, and it has been conjectured that he had been employed by Hawksbee[27]. The bill is undated. However, Helsham, who became a Fellow of Trinity in 1704, was Professor of Natural Philosophy from 1724 to 1738. The bill is for a variety of demonstration apparatus, such as "the diving experiment", "the mechanickall powers", "2 engin ioynts", "a grate screw to the iappand [japanned] fountain", and "the aeolipile". The total is not particularly large – £6-7-0 – and does not include any charges for transport. This last omission, and the presence of some items such as the engine joints and the screw, make one wonder whether Vream was working in Dublin.

Even more intriguing is a bill of 1729 for payment of £83-3-10 "for a fire Engine for the Use of Trin: Coll: Dub."[28]. Unfortunately it is not made out to an instrument maker, but to a

London firm of financial brokers, Messrs Newsham & Company. At almost the same time, the University of Glasgow bought a "fire Engine" from Jonathan Sisson the elder in London – none other than the model of a Newcomen engine which subsequently inspired James Watt to devise the separate condenser for the steam engine. If the Dublin engine was a model steam engine, then it must have been the only Newcomen engine ever to work in Ireland. Circumstances suggest that it is likely that Helsham would have wanted such an engine: he was an active teacher and Vream's bill indicates an extensive collection of demonstration apparatus. Four years before Helsham had been appointed to the chair, an inventory of philosophical instruments listed 79 items including a double barrel air pump with a large selection of accessories, a centripetal table, Culpeper's microscope, a magic lantern, a pneumatic fountain and Hero's fountain ("bruis'd & leeky")[29]. Helsham's lectures, summarised in his *Lectures on Natural Philosophy* (1739), ranged widely, and it is difficult to believe that he would be satisfied with the instruments which had been available in 1720. That Helsham had practical interests is clear from his help with improvements in Dublin's water supply, for which he was made an honorary freeman of the city[30]. His book contains discussions of practical problems, such as the motion of wheels over obstacles, which one would not expect to find in a course of natural philosophy[31].

In 1731 Stokes published *The Description and Uses of the Universal Aequinoctial Double-Ring Dial* (*Fig.* 3.2), and an example of this is in the National Museum of Ireland[32]. It appears to be identical to the instrument illustrated in Stokes' pamphlet, which makes it all the more likely that Stokes did indeed make his own instruments. Its design is entirely unremarkable. The market in which Stokes was selling his instruments is hinted at by the fact that the copy of his *Description* in the National Library of Ireland was presented by the Countess of Mayo[33]. Another instrument, a brass horary quadrant for latitude 52° 20′, signed and dated 1738, is also known[34]. A horizontal compass dial with folding gnomon, signed "Stokes fecit 1742" also survives[35]. At the end of his tract on the water supply of Dublin he included *The Mathematical Cabinet of the Hydrostatical Ballance Unlock'd: OR, an Easy Key to all its Uses* but, as with the "Pantometron", no examples are known to survive.

In 1711 Gabriel Stokes married Elizabeth King and they had two sons: John, who became Regius Professor of Greek and Archbishop King's Lecturer at Trinity College and was also grandfather of the famous nineteenth century scientist Sir George Gabriel Stokes FRS, President of the Royal Society – whom we will mention in Chapter Six; and Gabriel, who became Professor of Mathematics at Trinity and later Chancellor of Waterford Cathedral, and was the ancestor of a distinguished line of Dublin physicians. A number of the London instrument makers of the seventeenth century were comparatively rich – they were perhaps typical of those who traded in luxury goods in a wealthy city. For this reason sons of comparatively affluent men were from time to time attracted to the trade. At least one, Hilkiah Bedford, sent his son to university[36]. Stokes was unusual in that he was the only Dublin instrument maker known to have sent his sons to university before the second half of the nineteenth century.

Elizabeth King may well have been a relative of a mathematical instrument maker whose address was at Temple Bar, Dublin, between at least 1767 and 1772[37]. William King is known to have produced a brass sundial[38], a brass protractor[39] and a large mahogany octant with a boxwood scale, stamped "Made by Willm King Dublin for Mr John Iackson 1784", showing that he was active fully twelve years after he disappeared from the street directories[40]. It was probably a close relative of his – father or a brother – whose death was reported in the press at the end of 1771: "On Temple-bar, Mr Thomas King, Mathematical Instrument Maker"[41]. As the address was the same as that of William, it is possible that they worked together. William King advertised for an apprentice a few years previously, in which he revealed his own probable origins, which might have been guessed from his name: "A Protestant Lad, fourteen Years of Age, is wanted as an APPRENTICE. He must be well recommended, and the fee will be moderate. Enquire of Mr WILLIAM KING, Mathematical Instrument Maker on Temple-Bar"[42]. It is likely

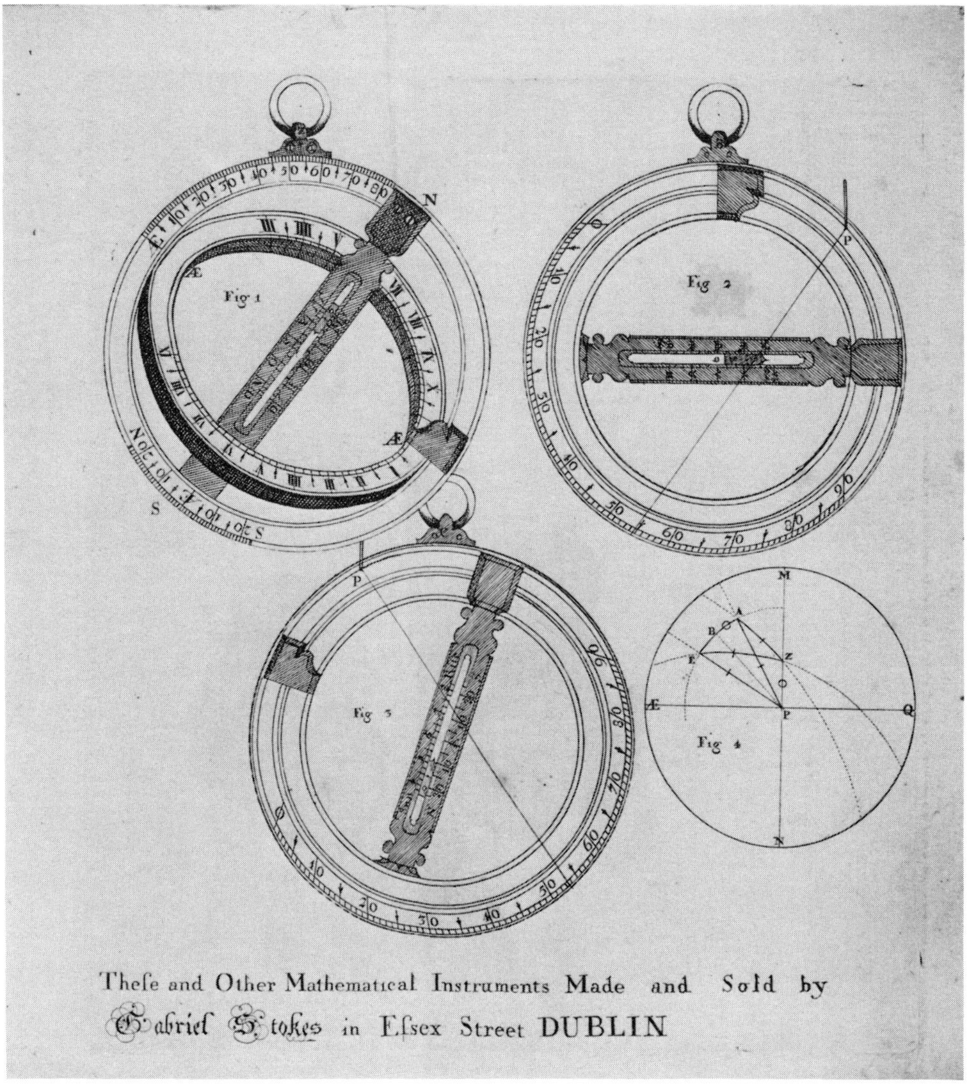

3.2. Plate from Gabriel Stokes, The Description and Uses of the AEquinoctial Ring-dial, *Dublin, 1731. National Library of Ireland.*

that most of the earlier instrument makers were Protestants, probably settlers from England (to judge from their names), and who may have learned their skills in London before removing to the second city.

Gabriel Stokes also collaborated with another surveyor who produced instruments, Thomas Cave; together they surveyed and produced a map of the site of the Parliament House in 1734[43]. Andrews states that both Cave and Stokes "produced equally versatile apprentices: Stokes' was Thomas Reading and Cave's was Jonathan Barker"[44]; neither made instruments. However, apparently William King claimed that he was Cave's only pupil, so perhaps he learned instrument making in Dublin[45]. Four circumferentors, each signed by Cave, are known (*Fig.* 3.3); also a pocket horizontal dial[46].

3.3. Surveyor's compass or circumferentor, signed "Thos. Cave + Dublin Fecit". National Museums and Galleries on Merseyside.

Surveying

In the seventeenth century, surveyors had concerned themselves little with the height of the land which they were surveying. The oldest levelling instruments were the water level, the plumb-bob level, and levelling staff, all known in classical Alexandria[47]. These instruments appear to have been little used in the seventeenth century: fairly accurate land surveying could be performed without measuring heights. An accurate knowledge of height, however, was necessary when the movement of water was being considered – that is, in questions of drainage, water supply and artificial navigation. Hydraulics represented the most important technology in the seventeenth and eighteenth centuries; it filtered into many areas of thought. Samuel Hartlib, born in Bohemia, an early Fellow of the Royal Society and an enthusiast for the new knowledge, wrote the Epistle Dedicatory for Gerard Boate's *Ireland's Naturall History* (1652). In referring to the limited interest in geography which he deplored, the expression he used was "the dispensation of wisdome hath not yet opened the conduit pipes of Natural Knowledge"[48].

The idea of a still-water navigation along an artificial cut was a product of the early eighteenth century. The Newry Navigation was built between 1731 and 1742 to transport coal from the area around Dungannon and Coalisland to Newry[49]. McCutcheon has pointed out that it was "the earliest true summit-level navigation of the canal era"[50], and that it predates both the Sankey Canal at St Helens and the Bridgewater Canal at Manchester[51]. The first proposals for a canal from Lough Neagh to Newry had been produced in the 1640s, and in 1703 Francis Nevil, "collector of her majesty's revenue", "survey'd and level'd" the ground[52]: he probably used a water or spirit level. Horner points out that Nevil's map is one of the earliest Irish maps to contain data on heights – and that his measurement of heights is not very accurate[53].

The Huguenot Richard Castle arrived in Ireland in 1728, and found employment in the office of Sir Edward Lovett Pearce, later Surveyor-General. The Newry Navigation got its Act of Parliament in 1729: Pearce was its engineer. Castle wrote an *Essay on Artificial Navigation*, which remains in manuscript in the National Library of Ireland[54], and which displays an excellent knowledge of contemporary continental practice. McCutcheon reasonably conjectures that Castle may have been responsible for much of the canal work before Pearce's death in 1733 – he certainly did oversee operations for the three seasons of 1734, 1735 and 1736[55]. Castle emphasised the need for an exact survey of levels[56]:

> The Survey and Level may be taken at the same time, by the means of an Instrument newly invented some time before I left England. I brought one of them over with me, which perhaps is the first of that kind brought into this kingdom.

Horner has suggested that Castle was referring to an altazimuth theodolite of the type devised by Jonathan Sisson of London[57]. Alternatively, it may have been the design produced by another London instrument maker, Thomas Heath. Sisson's instrument was first described in 1725 and Heath's in 1726[58]. An example of an early form of Heath's altazimuth theodolite survives[59] and it is quite clearly an instrument of considerable complexity, incorporating not only an altazimuth movement and divided arc, but also two telescopic sights. Sisson's instrument was quite similar. Castle had imported the most modern technology; indeed it is unlikely that any of the Dublin craftsmen of the 1720s would have been able to make such an instrument. Throughout Europe there was only a small demand for surveying instruments of this complexity in the eighteenth century[60], though a plan of Waterford (1764) by Scalé and Richards illustrates an altazimuth instrument in the cartouche of the dedication[61].

The Irish surveying instruments made in quantity between c.1775 and c.1825 were mostly circumferentors. There were three recognised difficulties in making circumferentors, according to contemporaries: irregular division of the scales, the compass needle not being mounted at the centre of the divisions, and the presence of iron in brass[62]. Bennett has produced a definition of the name, as some confusion has arisen:

> The circumferentor was a form of surveying compass – a magnetic compass with two fixed sights. The essential difference between the simple theodolite and the circumferentor is that bearings are indicated with the former by an alidade on a divided circle, with the latter by a magnetic needle (or, in the early forms, by a compass card mounted on the needle).

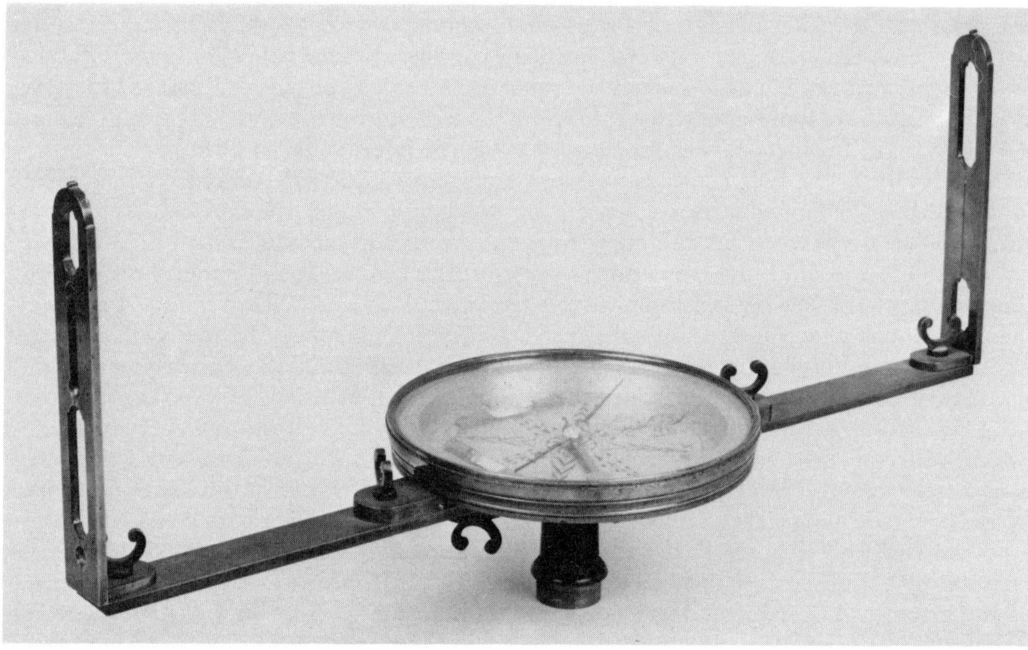

3.4. Surveyor's compass or circumferentor, signed "Seacombe Mason 6 Essex Bridge Dublin". Whipple Museum of the History of Science, Cambridge.

In the former case, the circle is kept fixed while the pointer (alidade) moves; in the latter, the circle moves while the pointer (magnetic needle) is stationary. This distinction was to become important in the development of different types of surveying instruments, suited to different working contexts. While the term "circumferentor" was consistently used to refer to a surveying compass, a standard form of instrument was not established until the eighteenth century[63].

When the standard form was achieved it proved to be most popular in America and Ireland: Bennett explains that this was because the landscape was uncluttered by man-made landmarks, unlike Britain and Europe, where the theodolite or plane table could reproduce the angles already established by fields and buildings[64]. There are discussions of the development of this instrument[65], and its use in America[66], as well as an assessment by Andrews of its long lifespan in Ireland[67].

The largest group of surviving instruments from the later eighteenth century are these surveying compasses or circumferentors, which all bear a distinct resemblance to each other, although they are engraved with the signatures of several different makers' – or retailers' – names. Examples by Lynch, Mason (*Fig.* 3.4), Spear, Spicer, Seward, and Walker are all known, and it is possible that they were all made by the same maker or engraved by the same engraver and were then sold over a fairly long period of time – some forty years – by different firms. Certainly, some of them have dates engraved on them, as well as their owner's names, the earliest being 1771 (sold by Spicer) and the latest 1807 (sold by Walker & Son)[68]. They were probably on sale for more than forty years. It is not surprising, given the range of activities covered by Stokes in the first half of the eighteenth century, and some of the Lynch family (see below) in the first half of the nineteenth, that there is evidence that some of these makers of surveying instruments also practised as surveyors. William Walker is one example[69]. Surveying was important for political and economic reasons: the decision to map Ireland at a scale of six inches to the mile was taken in 1724 by concerned members of the Dublin Parliament, determined that land values could be calculated from it. Benjamin Noble began the preface to his textbook on surveying, *Geodesia Hibernica* (1768), "Land is now become of so high a price in this kingdom, that the Art of Surveying ought to be performed with greater accuracy than perhaps is commonly the case"[70]. The Ordnance Survey's work in the 1820s was done exclusively by English military surveyors because, as the Marquess of Wellesley wrote to Lord Wellington, it "cannot be executed by Irish engineers and Irish agents of any description. Neither science, nor skill, nor diligence, nor discipline nor integrity sufficient for such a work can be found in Ireland"[71]. As Bennett has pointed out, "the military importance of the Irish survey was one factor in determining the level of detail"[72]. Another is that rents and absentee landlordism loom large in the political history of Ireland. It must also be borne in mind that there was a large agricultural population land-hungry for what meagre property was available. Dublin, as capital city, presumably supplied surveying instruments for the whole country.

Tradesmen and trade

Another section of the "professional" market as defined by Bryden and Bedini, was that of the navigator. Dublin's trade during this period was expanding, and meant that the port was large enough to sustain ships passing to and from London and Bristol. However, the number of nautical pieces which survive with the signatures of Irish instrument makers is considerably less than the quantity of circumferentors. This presents the possibility that the sailors were from the English rather than the Irish ports, and so preferred to buy their instruments at home. Nonetheless, some nautical instruments have been recorded, all of them wooden backstaves: one, now in a private collection, is signed "Gab. Stokes DVBLIN Fecit"; one dated 1733 by James Woodside, who died in 1743[73]; two by Alexander Stephens (*Fig.* 3.5), one of which is signed "Made by Alexander Stephens for Gyles Fearon 1747"[74]; and three by George McEvoy, one traditionally associated with Captain James Cook, another signed "Made by Geo. McEvoy, Temple Bar, Dublin for John Pearson 1743"[75]. All of these are extremely useful as evidence, for all of them date from before the run of street directories begin.

In the study of the London instrument making trade it has been shown that guild records are a major source for historical research[76]. This has not proved to be the case in Dublin. In the middle of the seventeenth century there were twenty-five guilds there[77], but scarcely any records survive. The one set of records which is almost complete is that of the Goldsmiths – and George Parker is the only eighteenth century Goldsmith who is known to have sold scientific instruments. From lists of voters it is possible to assign a few makers to guilds[78]. Two Edward Spicers were members of Sadlers Guild (presumably father and son, one or both instrument makers), Gabriel Stokes was a Cooper. Francis Booker voted in 1761, but he did so by right of being a freeholder, and no guild is recorded against his name. Two John Bookers voted: one in 1761, a Sadler; the other, in 1768, a member of the Trinity Guild.

Another example of a man with more than one interest who made instruments is George Parker. Parker, a member of the Goldsmiths' Company, is known to have been active as a watchmaker between 1704 – when he took his freedom – and his death in 1766[79]. Judging from the fact that he enrolled six apprentices between 1712 and 1728[80] he would seem to have employed several men. There is an equatorial ring dial signed "GEORGE PARKER DUBLIN" in a private collection[81]. The National Museum of Ireland has a watch signed "RINGLAND BELFAST" which is extremely similar to Parker's own timepieces: it is possible that he was selling some of his instruments wholesale. A different pattern of trade is illustrated by an advertisement issued in 1754 by the Round Glass House[82]. At the end of a long list of products were cited ". . . all sorts of Tubes, Globes, &c. for Electrical Experiments, Weather Glasses, Receivers for Air Pumps, and all Sorts of Philosophical Experiments . . .".

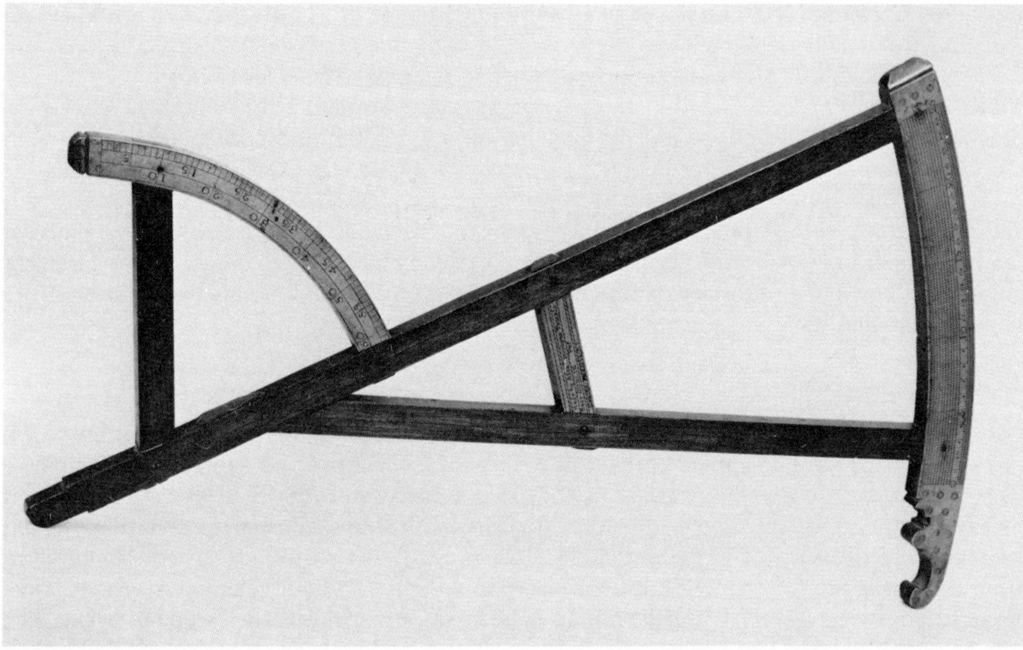

3.5. Backstaff, signed "Made by Alexr. Stephens Temple Barr DUBLIN For George Waddrington". Whipple Museum of the History of Science, Cambridge.

Just as in large centres of population in Britain, the Huguenots, and later Italians, brought their skills with them when they left their homeland, so Dublin's instrument trade benefited from the arrival of foreigners. John Margas's removal from London to Dublin is merely one example of the moving of a Huguenot instrument maker[83]; another highly talented Huguenot who was trained in London but carried out his most important work in Ireland was James Gandon, architect of many of Dublin's most notable buildings, including the Custom House (1781-91) where William Speer was Supervisor and Assayer of Spirits in the Port of Dublin, the Four Courts (1786-1802), in whose shadow stood Mason's shop on Arran Quay; and extensions to the Parliament House, which became the Bank of Ireland, and where Thomas Grubb was engineer. The same pattern of carvers, gilders, glassblowers and barometer makers springing from an Italian community seems as common to Dundee as to Dublin. For instance, barometers are known to have been produced – or at least signed – by James Del Vecchio, Alexander Feroni, Andrew and Joseph Gatty, Dominick Milessio and Anthony Pitzoli. Other barometers were sold by opticians who would respond to this growth in the "dilettante" market, which occurred in the later years of the eighteenth century[84]. And this does not exclude at least two indigenous "weatherglass makers" known to have been working in the mid-eighteenth century, Samuel Gregory and James Monaghan.

There is evidence that by the middle of the eighteenth century instruments were being imported from England, and that large firms who dealt in a wide range of goods were handling instruments as part of their trade. At Castletown, Co. Kildare, is a bill head of Thomas Jackson of Dublin, dated 28 November 1747:

Gentlemen and ladies may furnish themselves with variety of Looking Glasses, both in

Gilt & Plain Frames of the best and newest fashions, Crack Glasses, Bell Lamps, Shades, Globes for Halls, and variety of Gilt Branches.
N.B. He has lately imported from London a great choice of the best Crystal Spectacles, such as are Approved by the Royal Society. Green Spectacles, Reading Glasses, Concave Glasses, Telliscopes [sic], Opera Glasses, Pocket Microscopes, Pocket Weather Glasses, Concave Mirrours.

In many cases, as here, the mechanism was that a Dublin tradesman acted as middleman between a London wholesaler and the Irish customer. It is highly likely that, as a result of passing through an extra pair of hands, and of the costs of transport, the price in Dublin would be higher than the London price. Even when a large institution such as Trinity College dealt directly with a maker in London, the additional cost imposed by the remoteness of the metropolis could be appreciable. In 1793 Trinity paid off its debt to the agents of Jesse Ramsden. Over the nine preceeding years they had purchased a transit instrument, air pumps, and some second-hand instruments. The overall bill totalled £330-17-0. Of this, packing (including Ramsden's own supervision), insurance, and chairmen from Piccadilly to London docks, cost no less than £55-3-0[85]. The cost of shipping from London to Dublin was not included, nor the Customs dues payable at Dublin. Overall, the sum which Trinity paid above the actual cost of the instruments must have been about a quarter of their price.

It is difficult to assess the size of the market for instruments among the gentry, though there are indications that it was far smaller than in England. During the eighteenth century, itinerant philosophical lecturers travelled the length of England giving lecture-demonstrations, but there is little evidence of their activity in Ireland. Nor are advertisements known for the sale of instruments by subscription, such as are found in contemporary Edinburgh[86]. Such evidence as there is for the supply of "dilettante" instruments suggests that they were sold by merchants who were already dealing in luxury goods, such as the Round Glass House and Thomas Jackson. The nature of Irish society, too, was different from that in England. The landed class were in the eighteenth century almost exclusively "Anglo-Irish", descendants of English settlers who had been granted estates as native landlords were driven out. Reading their memoirs, it is hard to avoid the conclusion that many of them were more interested in their horses and their wine than in contemplating the complexities of nature.

Beauing, belling, dancing, drinking,
Breaking windows, damning, sinking,
Ever raking, never thinking,
Live the rakes of Mallow.[87]

The Lynch family

Members of the Lynch family are known to have been active as instrument makers between 1767 and 1846. Three generations appear to be involved; unfortunately there are three James Lynches, one in each generation. The eldest James Lynch is known from a single bill in Trinity College. Dated 1767, it is for "a new Air-pump, Gauge and Receivers" at a cost of £40[88]. James Lynch II died an old man in 1833. A variety of bills dating 1785 to 1825 all appear to be in the same handwriting – and in a different hand from the bill of 1767 mentioned above. The third generation was represented by James III and George; though the final name of the firm – Lynch & Co. – suggests that at least one other principal was involved at the end.

James Lynch II was assistant to the Professor of Natural Philosophy at Trinity. Records are not complete, but they imply that he was employed from c.1785 until his death, and that James Lynch III worked in the laboratory from 1833 to 1840[89]. James III's brother, George Houston Lynch, surveyed some of Trinity's lands in Co. Limerick in 1816[90], and was assistant in Humphrey Lloyd's magnetic observatory from 1838 to 1856[91]. As far as can be judged, James Lynch II was the leading supplier of experimental apparatus to Trinity during his lifetime. Before 1825 his "assistance" – presumably demonstration – earned him £30 a year, and after 1835 his son was receiving £60. With that in mind, they are known to have been paid £15 or more by Trinity for instruments in the following years[92]:

1793	£65	3	6	
1794	15	0	0	includes repairs
1797	16	0	4	includes work for Dunsink Observatory
1800	30	18	7	includes repairs
1802	38	1	4	
1803	15	15	0	
1804	19	4	1½	includes some Dunsink repairs
1808	20	0	0	"for his attention to the erection of the great circular instrument at Dunsink"
1809	24	15	7	
1815	26	2	8½	
1817	16	14	6	
1819	46	9	10½	includes some repairs
1820	60	0	0	"an electrical machine"
1821	81	11	5½	
1823	94	1	6	
1824	20	0	0	
1839	28	3	11	
1840	29	8	9	includes some repairs

The total of these payments is £629-1-10: clearly the Lynches derived an irregular but substantial income from Trinity. In addition to smaller payments not listed above, the incompleteness of existing records, and the practice of the Bursar making payments to professors for monies they had spent – sometimes sums of over £100 – makes it impossible for us to know how much was paid to Lynch for apparatus. A figure of £1,000, giving an average of £20 a year over the half-century from 1790, seems a reasonable estimate. The Lynches supplied a large number of comparatively inexpensive philosophical instruments to Trinity. Astronomical instruments, in contrast, could be costly. In 1808 Matthew Berge of London completed a mural circle for Dunsink Observatory, for which Trinity paid £964-9-6[93].

One bill is of particular interest – the earliest extant one for work by James Lynch II other than assistance with lectures. It covers work done between 1788 and 1793, and is endorsed by the Professor of Natural Philosophy, "The above work has been done in the Philosophy School & the charges appear to me reasonable"[94]. Although the bill includes a variety of repairs and alterations, it also lists a number of instruments which have been made, such as "a large horizontal dial for ye Garden". The role which Lynch was filling is similar to that of a modern laboratory technician.

From Lynch's work at Trinity arose another post: that of lecturer in natural philosophy to the Royal Dublin Society. The Royal Dublin Society, founded in 1731, had been modelled on the Royal Society of London. It differed from it, however, in placing a lesser emphasis on pure science and a greater one on subjects which might benefit the Irish economy, such as agriculture. It also saw itself as having an educational role, and to improve the effectiveness of its teaching a Professor of Chemistry was appointed in 1795. When the Royal Institution was founded in London in 1799, it began lectures in natural philosophy, and the Royal Dublin Society immediately took steps to follow this example. One of the Vice-Presidents, General

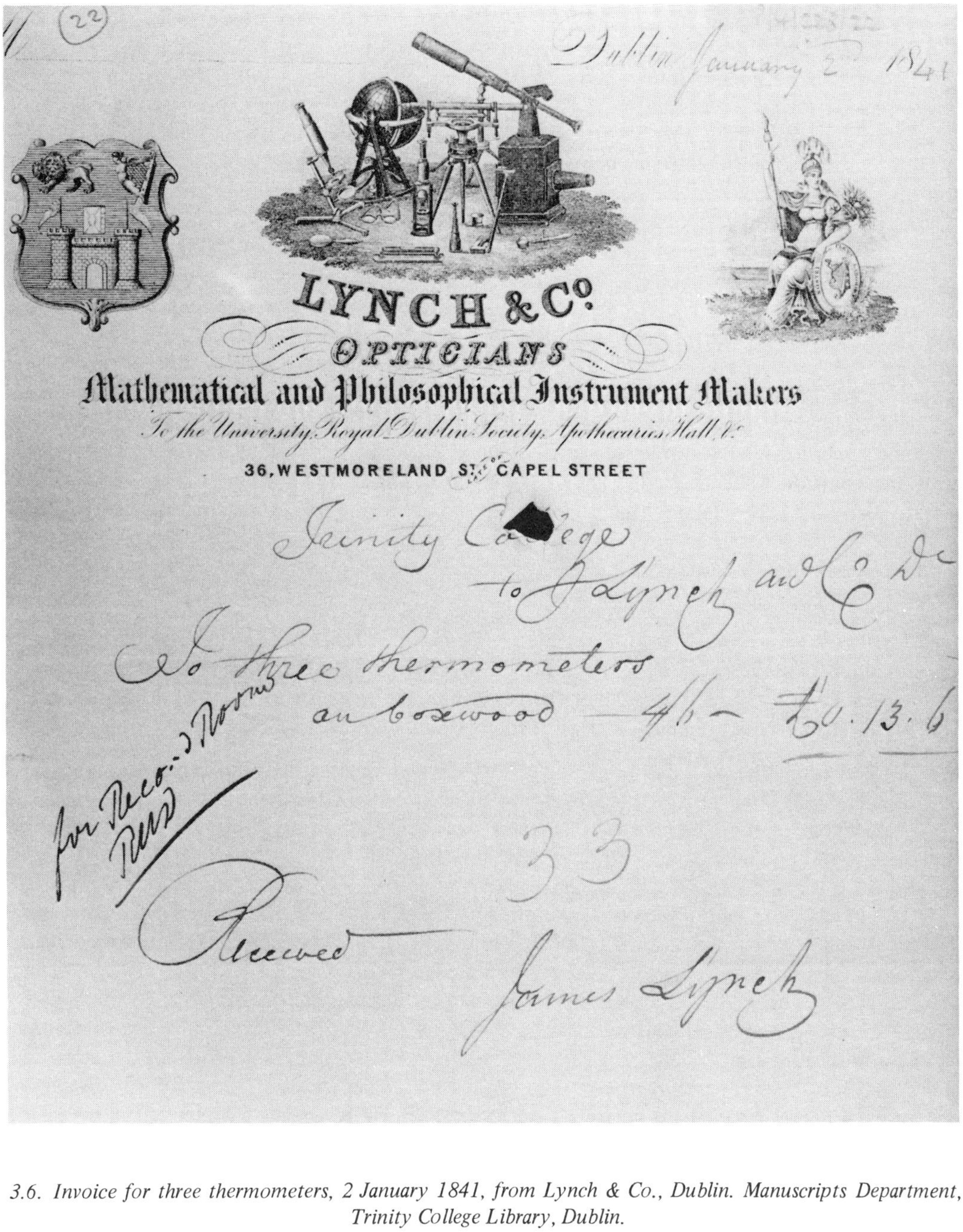

Dublin January 2nd 1841

LYNCH & Co.

OPTICIANS

Mathematical and Philosophical Instrument Makers

To the University, Royal Dublin Society, Apothecaries Hall &c.

36, WESTMORELAND St. CORNER OF CAPEL STREET

Trinity College

to J Lynch and Co Dr

To three thermometers on boxwood — 4/6 — £0.13.6

for Reco. Room

33

Received

James Lynch

3.6. *Invoice for three thermometers, 2 January 1841, from Lynch & Co., Dublin. Manuscripts Department, Trinity College Library, Dublin.*

Vallencey, was asked to "look out for a proper lecturer"[95]. The man he found was James Lynch. As a result of his connection with Trinity, he was able to present testimonials from Thomas Elrington, Erasmus Smith Professor of Natural Philosophy at Trinity, and his predecessor in that chair, Matthew Young, Bishop of Clonfert. Lynch agreed to give three courses of lectures each year for a fee of 75 guineas, plus £17-1-3 for keeping the Society's models in repair[96].

Two syllabuses for his lectures survive, and they show that he covered the whole of "natural and experimental philosophy"[97]. However, in 1815 he omitted astronomy, saying that he had previously lectured on it "under very unfavourable circumstances, for the want of an appropriate apparatus: and finding it extremely irksome to himself and uninteresting to his auditory . . . will decline entering upon that subject in the present course . . ."[98]. As a lecturer and instrument maker James Lynch II can thus be compared with figures in London such as Benjamin Martin, Frederick Accum and the Scotsman, James Ferguson.

Lynch wrote from time to time in the *Transactions* of the Royal Dublin Society: on the design of carriages[99], on a valve invented by Adam Oliver, model maker to the Society[100], and on improvements to the circumferentor suggested by an Irish-American, Richard Freeman[101]. Freeman's improvements were hardly revolutionary[102]. He suggested that a vertical scale be engraved on one of the vanes, so that angles of elevation and depression could be taken. As Lynch pointed out, this was not a new idea. Freeman also proposed that a vernier should be added to the horizontal scale. This offered only limited practical advantages, since Freeman's instrument did not have a telescope and probably could not have been made sufficiently mechanically rigid to enable angles to be measured to one minute of arc. The third suggestion made by Freeman was that the diurnal variation in magnetic declination be measured, using the vernier. Lynch accepted this idea, and reminded the Society that the importance of the diurnal variation to surveyors had been explained thirty five years previously by J. Hood of Derry. Lynch concluded his observations[103]:

> I will now take the liberty to mention a curious coincidence of idea between Mr. Freeman and myself on this subject [i.e. the diurnal variation], having for some time past projected a similar improvement; but, as I neither executed the instrument, nor published my intentions, Mr. Freeman remains unrivalled in his claim as the original inventor.

Lynch served the Royal Dublin Society for almost thirty-four years. By 1806 his salary had risen to £100[104] – though this appears to have included the previously separate payment for maintenance of models; by 1814 his salary was £200. This was two-thirds of the salaries of the three well-known men who were at the same time on the staff of the Society – William Higgins, Richard Griffith and Charles Giesecke[105].

Until 1831 the Dublin Society was funded partly by an annual grant from the government. When the grant was withdrawn, the Society found itself in a difficult financial position, and responded by cutting in half the salaries of its lecturer and its other professors. The Society seems to have been a little embarrassed by the action it had been forced to take since, when Lynch submitted a Memorial which discussed his own position, the Society had it printed[106]. Lynch pointed out that his salary was less than that of the other two professors, that he had lectured for no less than thirty-two years, and that he had himself procured the extensive apparatus which he used to illustrate his lectures. He ended by saying that the cut in his salary deprived him "in my old age, of my principal support, a great and manifest injustice after so many years' faithful service". It is difficult not to agree that he was treated shabbily.

Lynch presented his Memorial in December 1831; in May 1833 he died[107]. It then became clear that Lynch was going to be difficult to replace. He had given 75 lectures a year, repaired the Society's models, and attended from noon to 4 p.m. on one day each week to explain them – latterly for a salary of £100. The Society's Committee of Natural Philosophy now declared that this sum was inadequate for a suitable man, and added that a sum of between £200 and

3.7. Double barrelled air pump, signed "Lynch & Son, Dublin". Royal Dublin Society.

3.8. Armillary sphere, signed "Lynch & Son, Dublin". Private collection.

£500 would be needed to purchase suitable apparatus[108]. In the summer of 1833 John Stevelly of Belfast gave a course of lectures, for which he was paid £50, although had the Society been in an easier financial position it would have preferred to have given him more[109]. James Lynch III hired out his father's apparatus for £30[110]. By the end of January 1834 the Society had, with evident difficulty, found two candidates for the vacant lectureship. Testimonials were examined, committees consulted, two more candidates appeared, "probationary" lectures were given and, in the middle of April, the twenty-five year old Robert J. Kane was appointed[111]. Kane, already Professor of Chemistry at Apothecaries' Hall, Dublin, was to hold the lectureship until 1847. It was Lynch's fate to be succeeded by one of the greatest of Irish scientists, recipient of the Royal Medal of the Royal Society of London and a gold prize medal of the Royal Irish Academy.

James Lynch II is not known to have practised as a land surveyor, but in 1802 he proposed to make a map of Dublin County[112]. His business, on his own, and later with his son and other partners, lasted until 1846 and produced instruments mainly for the "dilettante" and teaching markets, judging by the surviving material. A double-barrelled pump now in the collection of the Royal Dublin Society (*Fig.* 3.7) may well have come from the Dublin Philosophical Society's collection. An armillary sphere is also known (*Fig.* 3.8) besides sundials, a large ring dial and such old-fashioned philosophical toys as a brass Culpeper microscope and a pocket globe[113]. James Lynch II was a maker, vendor and inventor of instruments, laboratory assistant and public lecturer. He was concerned with a variety of technical improvements. His son covered a similar range, including cartography and draughtsmanship[114] in place of lecturing and assisting in a laboratory. Yet it would be wrong to regard the Lynches as polymaths. Their activities, though varied, were related to one another, and in particular served the same group of customers. They were actors on a small stage, each having to play several parts.

During the nineteenth century, the instrument making and retailing trade in Dublin grew with new developments in science; descriptions in the directories demonstrate that these were followed by makers keen to keep their public up-to-date: Samuel Healy appears at the turn of the century as an "Electrifying Machine Maker". The most important businesses over this period were the family firms of Mason, Yeates and Grubb, the last of which assumed international status.

References and Notes for Chapter Three

1. D.J. Bryden, *Scottish scientific instrument-makers 1600-1900*, Edinburgh, 1972, 7.
2. S.A. Bedini, *Early American scientific instruments and their makers*, Washington, 1964, 3.
3. J.H. Andrews, *Plantation acres: a history of the Irish land surveyor*, Belfast, 1985.
4. Information about Gabriel Stokes is taken from a number of accounts: Sir William Stokes, *William Stokes*, London, 1898, 15-18; Joseph Larmor (ed.), *Memoir and Scientific Correspondence of the late Sir George Gabriel Stokes Bart.*, 2 vols, Cambridge, 1907, **1**, 1-2; Andrews, *op. cit.*, 271-273 (ref. 3).
5. Andrews, *op. cit.*, 272 n.27 (ref. 3) quotes a map of land near St Patrick Street, Dublin, St Patrick's Cathedral map collection, 17.
6. National Library of Ireland, 16.J.6(9).
7. Thomas H. Mason, Dublin Opticians and Instrument Makers, *Dublin Historical Record*, **6**, 1944, 138.
8. Trinity College Dublin, MUN/P/4/37/45.
9. Trinity College Dublin, MUN/P/4/37/46.
10. National Library of Ireland, 16.G.42(11).
11. J.H. Andrews, Introduction to George Taylor and Andrew Skinner, *Map of the roads of Ireland*, 2nd edn., Dublin, 1778; facsimile, Shannon, 1969, vii.
12. Andrews, *op. cit.*, 271, n.23 (ref. 3).
13. *Ibid.*, 272.
14. *Ibid.*, 273.
15. Offered for sale by Christie's New York 31 October 1985 Lot 237.
16. E.G.R. Taylor, *The mathematical practitioners of Hanoverian England*, Cambridge, 1966, 273, quoting from Robert Gibson, *A treatise of practical surveying*, 2nd edn., Dublin, 1763, 160. No copy of the first edition has been located, although it was advertised in the *Dublin Journal* of 9 March 1752.
17. J[ohn] A[nderson], Sinclair or Sinclar, George, in Sydney Lee (ed.), *Dictionary of national biography*, **52**, 1897, 293-294.
18. Whipple Museum, Cambridge, inventory no. 177: no provenance.

19. Trinity College Dublin, MUN/P/4/20/21.
20. K.T. Hoppen, *The common scientist in the seventeenth century; a study of the Dublin Philosophical Society 1683-1708*, London, 1970, 64.
21. St George Ashe to William Musgrove, 15 July 1687, Royal Society MS Early Letters, A.39, quoted by Hoppen, *op. cit.*, 64 (ref. 20).
22. Trinity College Dublin, *op. cit.*, (ref. 19). For a discussion of Hooke's instrument, see J.A. Bennett, *The divided circle: a history of instruments for astronomy, navigation and surveying*, Oxford, 1987, 67-72.
23. E.G.R. Taylor, *The mathematical practitioners of Tudor and Stuart England*, Cambridge, 1954, 240.
24. Rolf Loeber, *A biographical dictionary of architects in Ireland 1600-1720*, London, 1981, 31-39.
25. *Ibid.*, 36.
26. Trinity College Dublin, MUN/P/4/27a/41.
27. Taylor, *op. cit.*, 303 (ref. 23).
28. Trinity College Dublin, MUN/P/4/33/16.
29. Trinity College Dublin, MUN/P/1/669.
30. R.B. McDowell and D.A. Webb, *Trinity Colledge Dublin 1592-1952: an academic history*, Cambridge, 1982, 71.
31. Richard Helsham, *A course of lectures in natural philosophy*, edited by Bryan Robinson, Dublin, 1739, 369-375.
32. National Museum of Ireland, no inventory no. Ring dial: "Gab. Stokes Dublin fecit".
33. Shelfmark L01520.
34. National Maritime Museum, Gabb Collection F17. Signed: "Gab. Stokes Dublin Fecit 1738".
35. At the Willis Museum and Art Gallery Basingstoke, inventory no. BWM 1958.4.
36. Anthony Turner, A note on the life of Hilkiah Bedford, *Bulletin of the Scientific Instrument Society*, no. 9, 1986, 3-5.
37. *Walker's Dublin Directory*, 1767-1772.
38. Dublin Civic Museum: horizontal dial, brass, originally circular, cut down to octagonal shape, 5 inches across: "William King Dublin Fecit".
39. National Museum of Ireland: Brass protractor: "William King DUBLIN".
40. Private collection. Information from Peter Delehar, September 1986.
41. *Hibernian Journal*, 30 December 1771.
42. *The Freeman's Journal*, 29 November 1763, through December.
43. Mason, *op. cit.*, 139 (ref. 7).
44. Andrews, *op. cit.*, 272 (ref. 3).
45. Quoted *ibid.*, n.30, *Pue's Occurrences*, 29 September 1759 [not seen by authors].
46. The National Museums on Merseyside have one circumferentor: "Thomas Cave Dublin", inventory no. 8.2.1; another was offered for sale at Sotheby's 11 March 1963 Lot 115, signed "Cave, Dublin"; the third example was offered for sale by Historical Technology, *Catalog* 123, Spring 1982, item 154, signed "Thos. Cave * Dublin Fecit"; the fourth is in the National Museum of Ireland, signed "Thos: Cave Dublin Fecit". The pocket horizontal dial signed: "Tho^{S} Cave Fecit" is in the Whipple Museum Cambridge, inventory no. 1044.
47. Edmund R. Kiely, *Surveying instruments: their history and classroom use*, New York, 1947, 21-27.
48. Samuel Hartlib in Gerard Boate, *Ireland's naturall history*, London, 1652, sig. A3v-A4r.
49. W.A. McCutcheon, *The industrial archaeology of Northern Ireland*, Belfast, 1980, 52-55.
50. *Ibid.*, 49.
51. *Ibid.*, 54.
52. A.A. Horner, Some examples of the representation of height data on Irish maps before 1750, including an early use of the spot-heigh method, *Irish Geography*, **7**, 1974, 68-80.
53. *Ibid.*, 71.
54. National Library of Ireland, MS. 2737.
55. McCutcheon, *op. cit.*, 53-54 (ref. 49).
56. Castle, *op. cit.*, (ref. 54).
57. Horner, *op. cit.*, 71 (ref. 52).
58. J.A. Bennett and Olivia Brown, *The compleat surveyor*, Cambridge, 1982, 12.
59. Science Museum, inventory no. 1957-51, illustrated by Joyce Brown, *Mathematical instrument makers in the Grocers' Company 1688-1800*, London, 1979, plate 15; and an example of Heath's later "New Improved Theodolite" as described in J. Hammond, *The practical surveyor*, London, 1731, frontispiece, survives at Maynooth; this is illustrated in J. Fitzmaurice Mills, *The illustrated encyclopaedia of scientific instruments*, London, 1985, 83. Unfortunately, this piece lacks a provenance.
60. Bennett and Brown, *op. cit.*, 13 (ref. 58).
61. Benjamin Scalé and William Richards, A plan of the City and suburbs of Waterford, 1764.
62. Benjamin Noble, *Geodesia Hibernica*, Dublin, 1768, 37.
63. Bennett, *op. cit.*, 48-49 (ref. 22).
64. *Ibid.*, 149.
65. A. McConnell, An introduction to the surveyor's compass, *Land and Minerals Survey*, 5, 1987, 194-199.
66. Bedini, *op. cit.*, (ref. 2); D.J. Warner, The surveyor's compass, *Rittenhouse*, **1**, 1987, 64-75.
67. Andrews, *op. cit.*, 302-307 (ref. 3).
68. Leaving aside those surveyors' compasses which can be dated to the late seventeenth century (one signed by "W.R." and two signed by John Lewis) there are a few instruments which date from the early eighteenth century: these are examples by Gabriel Stokes (one sold by Christie's New York 31 October 1985 Lot 237),

Seward (one example noted by Dr Charles Mollan, another in the National Museum of Ireland, inventory no. 84-1940 (15)) and Thomas Cave (five examples: one noted by Dr Charles Mollan, a second in the National Museum of Ireland, a third in the National Museums on Merseyside inventory no. 8.2.83.1, a fourth sold by Sotheby's 11 March 1963 Lot 115 and a fifth offered by Historical Technology *Catalog* 123, Spring 1982 item 154). These pieces have some stylistic similarities, but there are enough individual idiosyncracies to enable one to believe that the signatory may have made the instrument. With the later eighteenth century circumferentors the similarities are greater. Examples recorded are: one signed by Barry of Cork, noted in *Bulletin of the Scientific Instrument Society*, no. 1, 1983, 7; one by Buckley in the National Museum of Ireland, inventory no. 2-1964; one signed by Wm. Davis offered for sale by Sotheby's 10 March 1987 Lot 132; four by Lynch, one in the National Museum of American History, a second in the National Museums of Scotland, inventory no. T1923.22, a third in a private collection and a fourth offered for sale by Sotheby's 13 November 1961 Lot 65 (part); two by Seacome Mason, one at the Whipple Museum of the History of Science, inventory no. 1899, another at St Patrick's College, Maynooth; two by T. & J. Mason, one at the Department of Engineering, Trinity College, Dublin, the other at the National Maritime Museum, Greenwich, inventory no. C.153; four signed Mason, one at the National Museum of Ireland, another at the University of Harvard, inventory no. 5158, a third offered for sale by Sotheby's 13 November 1961 Lot 65, the fourth offered by Sotheby's 11 June 1985 Lot 229; one signed Mason & Co. offered for sale by Christie's South Kensington 5 February 1987 Lot 75; five have been noted by Spicer, one in a private collection, a second at St Patrick's College, Maynooth, a third offered by Tesseract *Catalog* 10, Spring 1985 item 52 and again *Catalog* 21, Summer 1988 item 40, a fourth (dated 1771) offered by Sotheby's 5 March 1981 Lot 16, and a fifth by Sotheby's 25 February 1986 Lot 144; one signed by Spear & Clarke is at St Patrick's College, Maynooth; seven others are known by Spear, four in private collections, a fifth in Dublin Civic Museum, the sixth offered for sale by Sotheby's 31 March 1978 Lot 30, the seventh by Christie's South Kensington 14 December 1978 Lot 242; three by Walker, one (dated 1797) in a private collection, a second offered for sale by Phillips 4 March 1981 Lot 32 (part), the third (dated 1792) offered by Phillips 2 February 1984 Lot 84; seven signed by Walker & Son, one (dated 1807) in the National Museum of Ireland, two (one dated 1802) in a private collection, a fourth at St Patrick's College, Maynooth, a fifth in a private collection, a sixth offered for sale by Phillips 4 March 1981 Lot 32 (part) and the seventh by Sotheby's 11 June 1985 Lot 221.

69. William Walker, A map of that part of the lands of Ross, situate in the Parish of Linally, 1775, National Library of Ireland, 16.I.11 (12).
70. Noble, *op. cit.*, v (ref. 62).
71. Quoted by Bennett and Brown, *op. cit.*, 15 (ref. 58).
72. *Ibid.*
73. Science Museum, inventory no. 1912-393.
74. Whipple Museum of the History of Science, inventory no. 698; that dated 1747 is in a private collection, illustrated in D. Brieux, Marketplace, *Bulletin of the Scientific Instrument Society*, no. 19, 1988, 21.
75. One is in the collection of the Dick Institute, Kilmarnock; the example traditionally associated with James Cook is in the Museum of the Literary and Philosophical Society, Whitby; the dated example was noted by D.J. Bryden.
76. Brown, *op. cit.*, (ref. 59); M.A. Crawforth, Instrument making in the London guilds, *Annals of Science*, **44**, 1987, 319-377.
77. John T. Webb, *The guilds of Dublin*, Dublin, 1929, 252-256.
78. *An alphabetical list of the freemen and freeholders that polled at the election of members to represent the City of Dublin in Parliament*, Dublin, 1761; *Two alphabetical lists of the freeholders and freemen who voted in the late election of a Member to represent the City of Dublin in Parliament*, Dublin, 1768.
79. Geraldine Fennell, *A list of Irish watch and clock makers*, Dublin, 1963, 31.
80. H.F. Berry, The Goldsmiths' Company of Dublin, *Journal of the Cork Historical and Archaeological Society*, 2nd series, **8**, 1902, 29-50.
81. Described in Anon., *The Egestorff Collection an abridged catalogue*, n.p., n.d. [Dublin, 1982].
82. *Faulkner's Dublin Journal*, October 1754, reproduced in Mary Boydell, *Irish glass*, Dublin, 1976, inside front cover.
83. His father was the goldsmith Jacob Margas, discussed in T. Murdoch *et al.*, *The quiet conquest: the Huguenots 1685-1985*, London, 1985, 234.
84. N. Goodison, *English barometers 1680-1860*, 2nd edn., Woodbridge, 1977, 64-82.
85. Trinity College Dublin, MUN/P/4/67/24.
86. Bryden, *op. cit.*, 20-21 (ref. 1).
87. Eighteenth century song, author unknown.
88. Trinity College Dublin, MUN/P/4/55/31.
89. Trinity College Dublin, MUN/P/4/58/20 and

MUN/P/4/226/42.

90. Trinity College Dublin, MUN/P/4/157/14.
91. Trinity College Dublin, MUN/P/4/217/22 and MUN/P/4/261/42.
92. Trinity College Dublin, MUN/P/4/passim. Most of the accounts for 1825-1835 have been lost.
93. Trinity College Dublin, MUN/P/4/94/3, 3a, 3b. The Dunsink instruments ordered from London are discussed in greater detail by Patrick A. Wayman, *Dunsink Observatory 1785-1985*, Dublin, 1987.
94. Trinity College Dublin, MUN/P/4/68/16.
95. *Proceedings of the Royal Dublin Society*, **36**, 1799-1800, 121-122.
96. *Ibid.*, 133-135.
97. James Lynch, *Syllabus of a course of lectures in natural and experimental philosophy*. Two parts. Dublin, 1815.
98. *Ibid.*, second part, 11.
99. James Lynch, An essay on the construction and properties of wheel carriages, *Transactions of the Dublin Society*, **5**, 1806, 447-492.
100. James Lynch, Report: the double valve applied to the common pump, by Alan Oliver, model maker to the Dublin Society, *Transactions of the Dublin Society*, **5**, 1806, 447-492.
101. James Lynch, Observations on Mr Freeman's paper, *Transactions of the Dublin Society*, **6**, 1806-08, fifth pagination, 67-68.
102. Richard Freeman, On the improvement of the circumferentor, *Transactions of the Dublin Society*, **6**, 1806-08, fifth pagination, 47-66.
103. Lynch, *op. cit.*, 68 (ref. 97).
104. Expense of the Establishment for 1806, *Transactions of the Dublin Society*, **5**, 1806, appendix, xi-xii.
105. *The Dublin Society's account with the public commencing sixth January, 1814, and ending fifth January, 1915*, Dublin, 1815, 4.
106. A memorial from Mr James Lynch, *Proceedings of the Dublin Society*, **68**, 1831-32, v-vi.
107. *Ibid.*, **69**, 1832-33, 179.
108. *Ibid.*, 213-217.
109. *Ibid.*, **70**, 1833-34, 64-65.
110. *Ibid.*, 16-17.
111. *Ibid.*, 171.
112. Hely Dutton, *Observations on Mr Archer's Statistical Survey of the County of Dublin*, Dublin, 1802, 167.
113. The armillary sphere is in a private collection; one sundial was offered for sale by Phillips 5 April 1984 Lot 119, another dated 1824 was offered for sale by Sotheby's 23 October 1985 Lot 329; the large ring dial was offered for sale by Christie's New York 17 June 1986 Lot 60; the microscope, together with trade card in box, is in a private collection; the pocket globe dated 1813 was offered for sale by Sotheby's 27 May 1982 Lot 10.
114. Trinity College Dublin, MUN/P/4/172/17 and MUN/P/4/187/30.

CHAPTER FOUR

THE TRADE IN SCIENTIFIC INSTRUMENTS IN DUBLIN, 1830-1921

A.D. Morrison-Low

Mason, Yeates and Spear

Thomas H. Mason & Sons Ltd celebrated its bicentenary as a continuous family business in 1980, one of the few instrument firms to survive into the twentieth century[1]. Although Yeates & Son still have a shop in Grafton Arcade, the family are no longer connected with the business, which now consists of ophthalmic work only: Richard Spear's shop sign remains in Duke Street where he never had premises, as the sole unconnected reminder of his concern. However, in differing ways, these three firms proved to be the most prominent Dublin instrument makers dealing with the local Irish market during the nineteenth century.

The Mason family came originally from Seacombe in Cheshire, near Liverpool, and moved to Dublin as part of a government incentive in the early 1700s. Robert Mason, a tanner, became a Freeman of the City of Dublin in 1712, and was succeeded in this occupation by his son Thomas. The instrument business was established by Thomas Mason's eldest son Seacombe or Seacome Mason (1745-1804), whose family tradition had it that his fascination as a schoolboy watching the work of a French optician, who was a political refugee, led to an apprenticeship with Mason subsequently taking over the business when the Frenchman returned to his native country. However, the identification of this Frenchman remains a mystery, as John Alment, a possible candidate, died in 1787[2]: he had been foreman to another instrument maker with a French-sounding name, John Margas, who had in fact come to Dublin from London after his own business there had failed in 1758[3]. No matter with whom Seacombe Mason served his apprenticeship, he first appeared in the street directory for 1780 at No. 8 Arran Quay. The range of his wares is given in an advertisement which appeared in 1787[4]:

> . . . he has ready for Sale a great Variety of the following Articles, which he will sell on the most reasonable Terms, viz. Spectacles, Reading, Concave and Opera Glasses, Telescopes, Microscopes, Linen and Silk Provers, Barometers, Thermometers, Celestial and Terrestrial Globes of all sizes, Electrical Machines with Apparatus. Surveying and Drawing Instruments, Levels, Scales of Rules, Pantagrafts [sic], Goggles for preserving the Eyes from Dust or Wind, ditto for Children that Squint, Fowling Tackle, Oiled Silk and Linen Bathing Caps, Hat Covers, Aprons and Hoods for Riding, Umbrellas, ditto Repaired . . .

Perhaps the last items were a remainder from the old family business of tanning. A surveyor's compass with the signature of "Seacomb Mason 8 Arran Quay Dublin" is in the collection of St Patrick's College, Maynooth[5]. Mason continued trading until his death on 11 March 1804.

Two of Seacombe Mason's sons inherited the business, and ran it jointly for a few years at the Arran Quay address. They were Thomas (1781-1837) and Jonathan (1784-1849), and ran

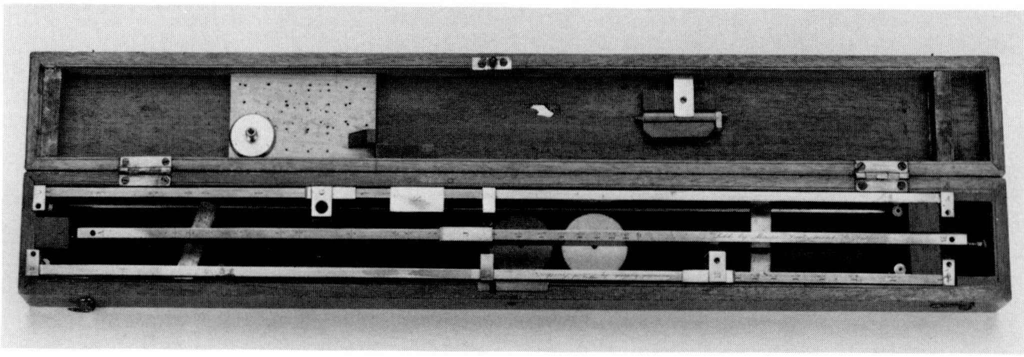

4.1. Copying device, sold by Mason, Parliament Street, Dublin. Trustees of the National Museums of Scotland.

separate ventures between 1809 and 1812, and then again after 1818; the business moved to Essex Bridge in 1809, where it remained for most of the nineteenth century (the street was renamed Parliament Street in 1883). Jonathan's second wife was from Limerick, and he eventually moved there, running an instrument business until his death in 1849[6]. Two items have been recorded with the signature "T. & J. Mason"; both are surveyor's compasses[7].

Thomas Mason continued the business in Dublin until his death in 1837. It has sometimes been claimed that he was responsible for "Mason's Hygrometer", invented by Dr James Apjohn, Professor of Chemistry at Trinity. In fact, the truth appears to be more convoluted: Apjohn certainly wrote about the theory behind the hygrometer[8], but the reason the instrument is known as "Mason's" is because of a paper by one John Abraham Mason, a surgeon from Pentonville, published in 1836, and acknowledging little of its precursors[9]. Apjohn did not mention the instrument making firm of Mason in his work, so that it can be surmised that they were not involved.

After the death of Thomas Mason, his eldest son Seacombe or Seacome (1808-1892) took over the running of the business. By his first wife, he had two children, Catherine and Thomas (1840-1912), and by his second wife, eight more, including a son named Seacome. The half-brothers worked in the firm with their father, but young Thomas was not happy either at home or at work. Finally he ran away to London where he managed to find employment with the large instrument making firm of Negretti & Zambra. He worked for them for a number of years.

Meanwhile, in Dublin, the Mason business went through some bad times. At one stage it looked as though the firm would have to be sold up, but Thomas Mason heard of this in London, returned to Dublin, secured a loan from his cousin John Holmes, and bought over the concern, taking over the management himself in 1887. He spent the remainder of his working life rebuilding the business and running it successfully, laying the foundations that have ensured its survival well into the twentieth century. He was a founder member of the Irish Optical Society, and brought back the optometer on his return to Dublin. He was also among the first in Ireland to sell equipment and chemicals for the new science of photography, which, with the invention of dry-plates, was becoming ever more accessible to the man in the street and not merely a hobby for wealthy amateurs. Mason's also began to sell the very popular magic lantern projectors, and lantern slides.

Thomas Mason died in 1913, and his son Thomas Holmes Mason (1877-1958), who had been working with the firm for a number of years, took over the management. He continued to expand the new premises in Dame Street, to which the firm had moved in 1894, and bought the adjoining property in 1914. This allowed the photographic department to grow; Mason's established a developing and printing works, to which Thomas H. Mason added a postcard printing department in 1932, which only survived into the early 1940s due to the rise in the cost of silver salts. He was himself a keen amateur photographer.

By the 1920s there were two distinct departments to the Mason enterprise; the photographic department, and the optics department, which was the main branch of the business. This included an ophthalmic optician's practice, but also sold the latest in laboratory equipment, as well as surveying instruments, and other scientific apparatus. After the First World War, Thomas H. Mason expanded the laboratory supplies part of the business into a separate department. He was joined at work by three of his sons, Standish, Barry and Alex. The firm became a limited company in 1935, with three departments: laboratory, photographic and optics, although the two latter were closed down after the death of Standish Mason in 1969. The laboratory supply side continues to expand today, and there are still members of the Mason family involved in the firm[10].

Another family business which began about the same time as the Mason firm, and continues today although no members of the original family are involved in what has become a purely ophthalmic optician's practice, was that of Samuel Yeates. He first appeared in the 1790 directory as an optician at Upper Ormond Quay, although the surname appears as early as 1769, with members of the family variously described as weavers, drapers and shoemakers; later there were brass founders, cutlers and surgical instrument makers. Thomas H. Mason wrote that he understood that "the first optician Yeates served his apprenticeship to Seacombe Mason, who married Anne Yeates in 1775. The families intermarried in more than one generation, and today [i.e. in 1944] there is still a Mason Yeates"[11]. In 1842 Seacombe Mason married Charlotte Augusta Yeates as his second wife[12]. As the Yeates family was so large, it would be useful to point out some of the more obvious relationships between the optical instrument makers[13].

Samuel Yeates (1762-1834) was the second child of Andrew and Mary Yeates, and was an optician from 1790 until his death; the firm became Samuel Yeates & Son in 1836. Samuel's younger brother Kendrick Yeates (1776-1850) owned a brass foundry, but sometimes did some

4.2. Trade card for Yeates & Son, Dublin, 1834-1839. Trustees of the Science Museum, London.

work as an optician and instrument maker; he married Margaret Faulkner in 1802 and they had seventeen children; however, none of them became instrument makers. In 1794 Samuel Yeates married Mary Ellen Yeates (1768-1840), daughter of Francis and Anne Yeates of Moorechurch, and they had nine children, all born at 29 Capel Street, where the firm was located between 1795 and 1810. George Yeates (1796-1862) was their second son (the eldest died while a child); he continued the business, and we shall deal with his family below. The third son, Andrew (1800-1876) became involved in the work when Thomas Drummond approached Samuel Yeates to construct the limelight which he had developed for the Trigonometrical Survey of Ireland; Andrew's enthusiasm for the precision instruments used in survey work dates from this time. He left for England in 1821, "and on his arrival in London he fortunately made the acquaintance of [Edward] Troughton, with whom he worked for several years. In 1833 he was employed at the [Royal] Observatory, under the superintendence of the great optician, for the readjustment and repair of the instruments"[14].

Andrew Yeates became a Fellow of the Royal Microscopical Society, was a Member of the Society of Arts, and a Fellow of the Royal Astronomical Society, to whose *Monthly Notices* he contributed a paper on the construction of transit instruments[15]. He was in business as an optician at 12 Brighton Place, New Kent Road, London, between 1837 and 1873, when he retired through ill health[16]. Of instruments with his signature, a prismatic compass, a two-foot Gunter's rule and a transit theodolite have been recorded[17]. He married twice, and his second wife was Elizabeth Varley, daughter of Cornelius Varley the instrument maker and designer[18]. He died at Buckhurst Hill, London in 1876.

A fourth son, Thomas (1802-1874), was also an optician; his career is not so well documented as those of his brothers. In 1846 he married Martha, the daughter of Robert Taylor, an optician at 193 Britain Street, Booterstown; they had no children.

The second son, George, was married three times, and had ten children by his first wife; the seventh child and second surviving son, Stephen Mitchell Yeates (1832-1901) inherited the family business of Yeates & Son and appears to have run it from about 1865 onwards. His younger brother Horatio (1834-1907) was also involved in the instrument trade: he exhibited a large induction coil at the Special Loan Collection Exhibition at South Kensington in 1876[19], and two instruments signed by him have been noted: a theodolite and a microscope[20]. Horatio Yeates was in business in Dublin for a very short time; he moved to London in 1864, where he was at 33 King Street until 1880 when he emigrated to South Australia. He appears to have continued to work as an optician in Melbourne and Hobart[21]. He and his wife had six children, of whom one, the youngest, Charles Samuel Sparks Yeates, worked at the Perth Observatory from 1898.

Stephen Mitchell Yeates had six children, but it is unclear whether Arthur Mitchell Yeates, the only son to survive his father, carried on the business. Stephen Mitchell Yeates' only sister to marry, Rebecca, had a large family of seven extremely bright sons and two daughters. She married a soap manufacturer, George Dixon, and their fourth son, William Vickers Dixon, completed seven years' apprenticeship with Yeates & Son between 1878 and 1885; he was appointed manager of their workshops, a position he held for the next three years. During his apprenticeship he took his degree at Trinity College, Dublin in 1882, obtaining a gold medal in his senior moderator examination. He later became Chief Inspector of Technical Schools in Ireland.

Yeates & Son made and sold a great range of scientific instruments, but the evidence for this is widely scattered and fragmentary[22]. By the mid-nineteenth century much of their material was aimed at the growing market for educational apparatus: with an increased demand for scientific apparatus in schools and colleges, there was scope for supplying reasonably-priced equipment locally. From 1852 until 1910 Yeates & Son advertised as "Instrument Makers to the University"[23], which implies that Trinity College was a good customer. George Yeates sold chemical apparatus to the university in 1838[24], and other sales are recorded between 1811 and

4.3. Electro-magnetic switch, signed "Yeates & Son, Dublin". University College, Cork.

4.4. S.M. Yeates' improved form of G.J. Stoney's local heliostat. Trustees of the National Museums of Scotland.

1857[25]. In 1827 a bill head advertised "All manner of CHEMICAL APPARATUS Made & repaired"[26]; the range on offer was broad, and the workmanship clearly versatile. At the end of the Crimean War, the firm was paid for "Work on illuminations to celebrate the peace with Russia".

A handbill dating from between 1847 and 1857 provides a valuable indication of the scope of Yeates' activities (*Fig.* 4.5)[27]. It states that they "Have constantly on hand the following articles, either selected from the best foreign makers, or manufactured on their own premises", and then gives quite a formidable list. Near the bottom of the sheet is the announcement, "The

YEATES AND SON,

2 GRAFTON STREET, DUBLIN,

OPTICIANS BY SPECIAL APPOINTMENT TO THE UNIVERSITY AND THE CORPORATION OF THE PORT OF DUBLIN,

Have constantly on hand the following articles, either selected from the best foreign makers, or manufactured on their own premises:—

Apparatus for Illustrating the newest discoveries in Acoustics.

Binocular Field Glasses.
Opera Glasses.
Spectacles of every description,
Best Brazilian Pebble Spectacles.
Railway Goggles.
Achromatic Microscopes.
Binocular Microscopes.
Microscopic Apparatus, and every requisite for Mounting Microscopic Objects.
Map Glasses and Codington Lenses.
Pocket Magnifiers.
Magic Lanterns
Stereoscopes.
Telescopes of every description.
Telescope Stands.
Wheel, Pediment, and Standard Barometers.
Aneroid Barometers.
Six's Self-Registering Thermometers.
Rutherford's do. do.
Bath Thermometers.
Chemical do.
Greenhouse do.
Daniel's Hygrometers.
Mason's do.
Anemometers.
Rain Gauges,
Drawing Instruments.
T Squares and Straight Edges.

Yeates and Son's ONE SHILLING MICROSCOPE, magnifying one hundred times. Post free for 13 stamps.

Parallel Rules.
Protractors.
Pentographs.
Tape Measures, and Standard Measures of every description.
Surveying Chains.
Levels and Theodolites.
Transit Instruments.
Sextants and Optic Squares.
Prismatic Compasses.
Station Staves.
Magnetic Compasses.
Electrical Machines and Apparatus.
Galvanic and Magnetic Apparatus.
Prometheans for the Coast Guard Service.
Sun Dials constructed with accuracy for every latitude.
Agents for Grautoff's Pressure and Vacuum Gauges.
Agents for Smith & Beck's Microscopes.
Air Pumps and Pneumatic Apparatus.
Polarizing Apparatus.
Spectroscopes and Pure Salts.
Electric and Magnetic Telegraphs.
Electric Bells.
Kaleidoscope for adapting to the Magic Lantern.
Apparatus for the rapid production of Ice for Domestic Purposes.

Yeates & Son's new £5 5s. set of Standard Meteorological Instruments, verified by Kew or Greenwich.

YEATES and SON have much pleasure in calling attention to the following Instruments recently Improved or Invented by them:—

Improved Equatorial Telescope Stand.
Do. do. especially adapted for Amateur Astronomers.
Large Dial Barometers, (Dial 3 feet, 6 inches diameter) suitable for Public Buildings.
Large Dial Thermometers, (Dial 3 feet, 6 inches diameter,) to correspond with Barometers.
New Elliptograph for describing Ellipses correctly in all proportions.
New Field Microscope, to fit in the waistcoat pocket; much recommended for Microscopists and Tourists.
Improved Double-acting Single-barrelled Air-pump.
Improved Oxy-Calcium and Oxy-Hydrogen Apparatus, for Dissolving Views and Microscopes.
Newly Improved Binocular Field Glass.
Newly Improved Binocular Opera Glasses.

THE TRADE SUPPLIED ON LIBERAL TERMS.

N.B.—In consequence of the loss arising from the difficulty of collecting small accounts for orders and jobs forwarded by post, YEATES & SON *beg respectfully to state, that for the future they cannot send jobs unless paid for previously, nor attend to small orders by post, except accompanied by a remittance.*

4.5. Yeates & Son handbill, dating from between 1847 and 1857. Royal Dublin Society.

trade supplied on liberal terms", making it abundantly clear that at this date Yeates were manufacturers in their own right[28].

Insight into the nature of the Yeates' business around 1880 is provided by a series of four catalogues which they issued between 1877 and 1887. They list over 2,000 items and Yeates probably made the majority of them. Only in two instances do they suggest otherwise, listing measuring tapes and chains by Chesterman of Sheffield[29], and microscope object glasses "of the best *English* make"[30]. Other apparatus was almost certainly imported: Gramme dynamos, for example[31], and some, such as their astronomical telescopes on equatorial mountings[32] must have been made elsewhere in Dublin – in this case, by Grubb. Some of the surveying instruments may have been made by another member of the family – Andrew Yeates of 12 Brighton Place, New Kent Road, London. His name appears in these catalogues on woodcuts of a portable theodolite and an improved telescopic prismatic compass[33]. Alternatively, he may have been only their designer – the names of H[oratio] and S[tephen] M. Yeates also appear regularly in this role.

Instruments were also playing a larger part in the practice of engineering, and engineering was itself expanding very rapidly. For example, the first railway in Ireland opened in 1834 and by 1894 there was a route mileage of 3,044[34]. Many of the early railway bridges, such as the Craigmore Viaduct (1851-1852), the Shannon Bridge at Athlone (1850-1851), and the Boyne Viaduct at Drogheda (opened 1855) were very large structures which required detailed preliminary surveying. Most of the engineers who built the Irish railways were Irishmen, many educated in the School of Engineering which Trinity established in 1841 – the second university in the world to do so[35]. The School had a "large supply of instruments" for the students' use[36]. The first Professor of Engineering, Sir John Macneill, took apprentices in his private practice. In addition to a premium of £150, he required that each apprentice should supply himself with a level and theodolite[37]. George Yeates made a number of attempts to improve surveying instruments. In 1845 he designed a theodolite which would be lighter than those in use, but equally accurate[38]. Then he devised a simple form of clinometer[39], a modified dumpy level[40] and a basic rangefinder[41]. One of the preoccupations of civil engineers working in Ireland was the drainage of water, and Yeates produced his own form of drainage level[42]. In all these instruments, Yeates was attempting to alter existing equipment so that it would be easier to use.

A large proportion of the entries in the Yeates' catalogues is for demonstration and teaching apparatus – such as the "Brass Boiler and Lamp on Carriage, to show the reactive force of steam"[43], and they also sold various teaching models, such as one illustrating the polarisation of light[44]. Yeates also sold apparatus for telegraphy, meteorology and some instruments for medical diagnosis, such as ophthalmoscopes and laryngoscopes[45]. At one end of their range we find walking stick telescopes and "charm microscopes" (to be worn round the neck)[46]; at the other, self-recording rain gauges and tide gauges[47]. Given the expansion of higher education in Ireland from the 1840s onwards, there was a significant market for educational material. A list of apparatus owned by the Royal Dublin Society, which was compiled about 1850, demonstrates that under Robert Kane their collection of apparatus had expanded[48]. The list describes 130 types of instrument or model, including models of a railway locomotive, Lord Rosse's telescope, and the "Atmospheric Rail Road"[49].

Whilst it would be possible for expensive and unusual items to be made in London or Birmingham, it is more likely that they would have been constructed in Dublin, where the purchaser could more easily explain his requirements and oversee the work. Thomas H. Mason discussed whether instruments could in general be economically made in Ireland:

> In olden times the opticians made a much greater proportion of the articles they sold. Today [i.e. 1944] many of them make nothing. Some who, like ourselves and Yeates, manufactured a number of various instruments have found it unpracticable to do so to-

day. Yeates made the last big attempt about 1890, but they were forced to market most of their products through the medium of English firms and they lost money. The home market is too small, and because of the advance of science the variety is too great. A world market is needed, and this requires not merely highly-educated scientific brains, but also large organising ability and great capital resources. One of my sons went through the works of the largest English makers of engineering instruments about 4 years ago. He saw rows of theodolites, each valued for about £100, and was informed that they were manufactured in batches of not less than 50 instruments; a smaller number was not economic. Fifty of these instruments would be sufficient for 25 years' demand in Eire[50].

The Yeates family were primarily makers and sellers of scientific instruments. They did, however, fill other roles in Dublin's scientific life. George Yeates published a meteorological journal annually for the Royal Irish Academy between 1843 and 1849[51], and contributed a note on a meteor to the Academy's *Proceedings*[52]. Stephen M. Yeates wrote on the use of a barometer to measure heights[53]. One of the family – almost certainly Stephen – attended some of the meetings of the Dublin Microscopical Club. This was a highly selective body. The total number of full members allowed by its constitution was no more than twelve: meetings were held in rotation in their homes. Each meeting had the same structure: the results of original investigations were presented first, then, "that which is not quite novel, and then 'pretty objects' should be shown when all the general business is over"[54]. In April 1866 "Mr Yeates exhibited a soap-bubble under the microscope, forming a very gorgeous object, owing to the magnificent and ever-changing play of colours which was presented"[55]. On other occasions Yeates showed to meetings a new design of microscope made by Smith & Beck of London, and a new pocket microscope[56].

S.M. Yeates became involved in some of the earliest experiments with the telephone in Ireland. Almost ten years before Alexander Graham Bell produced his system, a German, Phillip Reis, had produced a form of transmitter, and in 1865 Yeates demonstrated his modified version of this before the Dublin Philosophical Society "when both singing and distinct articulation of several words were heard through it, and the differences between the speakers' voices clearly heard"[57].

The third prominent nineteenth century Dublin instrument maker does not appear to have involved himself, as did the Yeates, in local scientific activities; Richard Spear seems to have written no papers. However, his shop was in an extremely prestigious position, both in Capel Street where he was first recorded in 1791, and in College Green where the firm was located until 1864 (*Figs* 4.6 & 4.7). Yet from 1818, it would appear that the instrument business was not sufficient to be financially viable on its own, as it was supplemented by trade in wax, oil and spermaceti candle manufacturing. A relative of Spear, Saunderson Spear (possibly Richard's son), subsequently appears to have run this side of the venture from various separate addresses before the firm changed its name to "Spears & Co." in 1838.

Little is known about the structure of any of these three businesses or how they operated on a day-to-day basis; but at least we have some idea of the trade succession in the cases of the Masons and the Yeates. With Spear, the evidence is too fragmentary. The most enduring part of Spear's business is his trade sign, a seaman in traditional costume holding an octant, which is still to be found in Duke Street[58]; Spear used it to head his newspaper advertisements and, when Charles Dickens visited Dublin to deliver one of his readings at the Rotunda, he went into the shop to enquire about the figure[59]. Dickens describes such a figure in *Dombey and Son*[60]; this is quoted by Bedini, who explains the popularity of these shop signs in the British Isles and America during the nineteenth century[61]. It is difficult to assess just how much of Spear's stock was manufactured as opposed to bought-in and retailed. He certainly was proud to announce that he could offer London-made stock[62]:

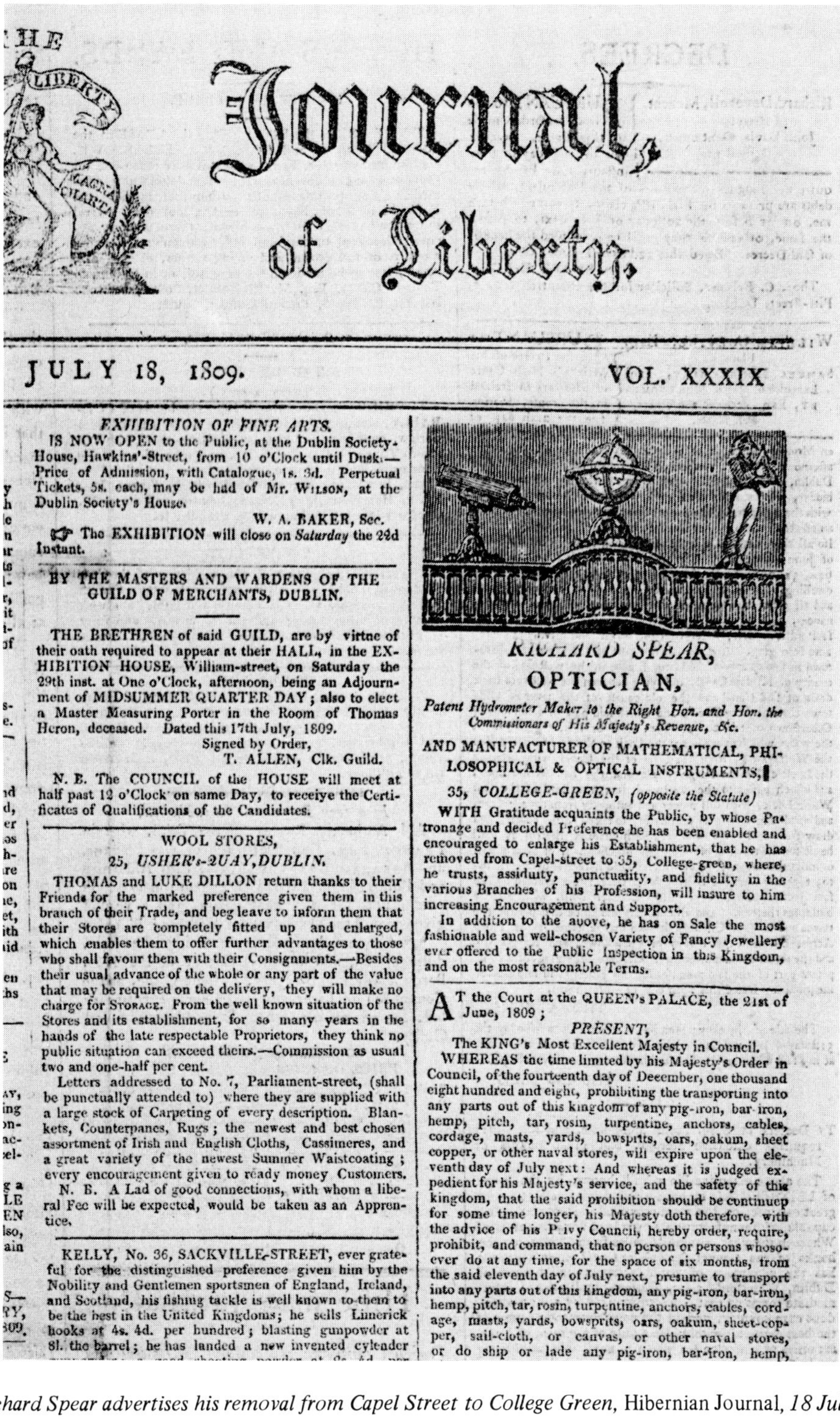

HE Journal, of Liberty.

JULY 18, 1809. VOL. XXXIX

EXHIBITION OF FINE ARTS.

IS NOW OPEN to the Public, at the Dublin Society-House, Hawkins'-Street, from 10 o'Clock until Dusk.—Price of Admission, with Catalogue, 1s. 3d. Perpetual Tickets, 5s. each, may be had of Mr. WILSON, at the Dublin Society's House.

W. A. BAKER, Sec.

☞ The EXHIBITION will close on *Saturday* the 22d Instant.

BY THE MASTERS AND WARDENS OF THE GUILD OF MERCHANTS, DUBLIN.

THE BRETHREN of said GUILD, are by virtne of their oath required to appear at their HALL, in the EXHIBITION HOUSE, William-street, on Saturday the 29th inst. at One o'Clock, afternoon, being an Adjournment of MIDSUMMER QUARTER DAY; also to elect a Master Measuring Porter in the Room of Thomas Heron, deceased. Dated this 17th July, 1809.

Signed by Order,
T. ALLEN, Clk. Guild.

N. B. The COUNCIL of the HOUSE will meet at half past 12 o'Clock on same Day, to receiye the Certificates of Qualifications of the Candidates.

WOOL STORES,
25, USHER's-2UAY, DUBLIN.

THOMAS and LUKE DILLON return thanks to their Friends for the marked preference given them in this brauch of their Trade, and beg leave to inform them that their Stores are completely fitted up and enlarged, which enables them to offer further advantages to those who shall favour them with their Consignments.—Besides their usual advance of the whole or any part of the value that may be required on the delivery, they will make no charge for STORAGE. From the well known situation of the Stores and its establishment, for so many years in the hands of the late respectable Proprietors, they think no public situation can exceed theirs.—Commission as usual two and one-half per cent.

Letters addressed to No. 7, Parliament-street, (shall be punctually attended to) where they are supplied with a large stock of Carpeting of every description. Blankets, Counterpanes, Rugs; the newest and best chosen assortment of Irish and English Cloths, Cassimeres, and a great variety of the newest Summer Waistcoating; every encouragement given to ready money Customers.

N. B. A Lad of good connections, with whom a liberal Fee will be expected, would be taken as an Apprentice.

KELLY, No. 36, SACKVILLE-STREET, ever grateful for the distinguished preference given him by the Nobility and Gentlemen sportsmen of England, Ireland, and Scotland, his fishing tackle is well known to them to be the best in the United Kingdoms; he sells Limerick hooks at 4s. 4d. per hundred; blasting gunpowder at 8l. the barrel; he has landed a new invented cylender

RICHARD SPEAR,

OPTICIAN,

Patent Hydrometer Maker to the Right Hon. and Hon. the Commissioners of His Majesty's Revenue, &c.

AND MANUFACTURER OF MATHEMATICAL, PHILOSOPHICAL & OPTICAL INSTRUMENTS,

35, *COLLEGE-GREEN, (opposite the Statute)*

WITH Gratitude acquaints the Public, by whose Patronage and decided Preference he has been enabled and encouraged to enlarge his Establishment, that he has removed from Capel-street to 35, College-green, where, he trusts, assiduity, punctuality, and fidelity in the various Branches of his Profession, will insure to him increasing Encouragement and Support.

In addition to the above, he has on Sale the most fashionable and well-chosen Variety of Fancy Jewellery ever offered to the Public Inspection in this Kingdom, and on the most reasonable Terms.

AT the Court at the QUEEN's PALACE, the 21st of June, 1809;

PRESENT,

The KING's Most Excellent Majesty in Council.

WHEREAS the time limited by his Majesty's Order in Council, of the fourteenth day of December, one thousand eight hundred and eight, prohibiting the transporting into any parts out of this kingdom of any pig-iron, bar-iron, hemp, pitch, tar, rosin, turpentine, anchors, cables, cordage, masts, yards, bowsprits, oars, oakum, sheet copper, or other naval stores, will expire upon the eleventh day of July next: And whereas it is judged expedient for his Majesty's service, and the safety of this kingdom, that the said prohibition should be continuep for some time longer, his Majesty doth therefore, with the advice of his Privy Council, hereby order, require, prohibit, and command, that no person or persons whosoever do at any time, for the space of six months, from the said eleventh day of July next, presume to transport into any parts out of this kingdom, any pig-iron, bar-iron, hemp, pitch, tar, rosin, turpentine, anchors, cables, cordage, masts, yards, bowsprits, oars, oakum, sheet-copper, sail-cloth, or canvas, or other naval stores, or do ship or lade any pig-iron, bar-iron, hemp,

4.6. Richard Spear advertises his removal from Capel Street to College Green, Hibernian Journal, *18 July 1809. National Library of Ireland.*

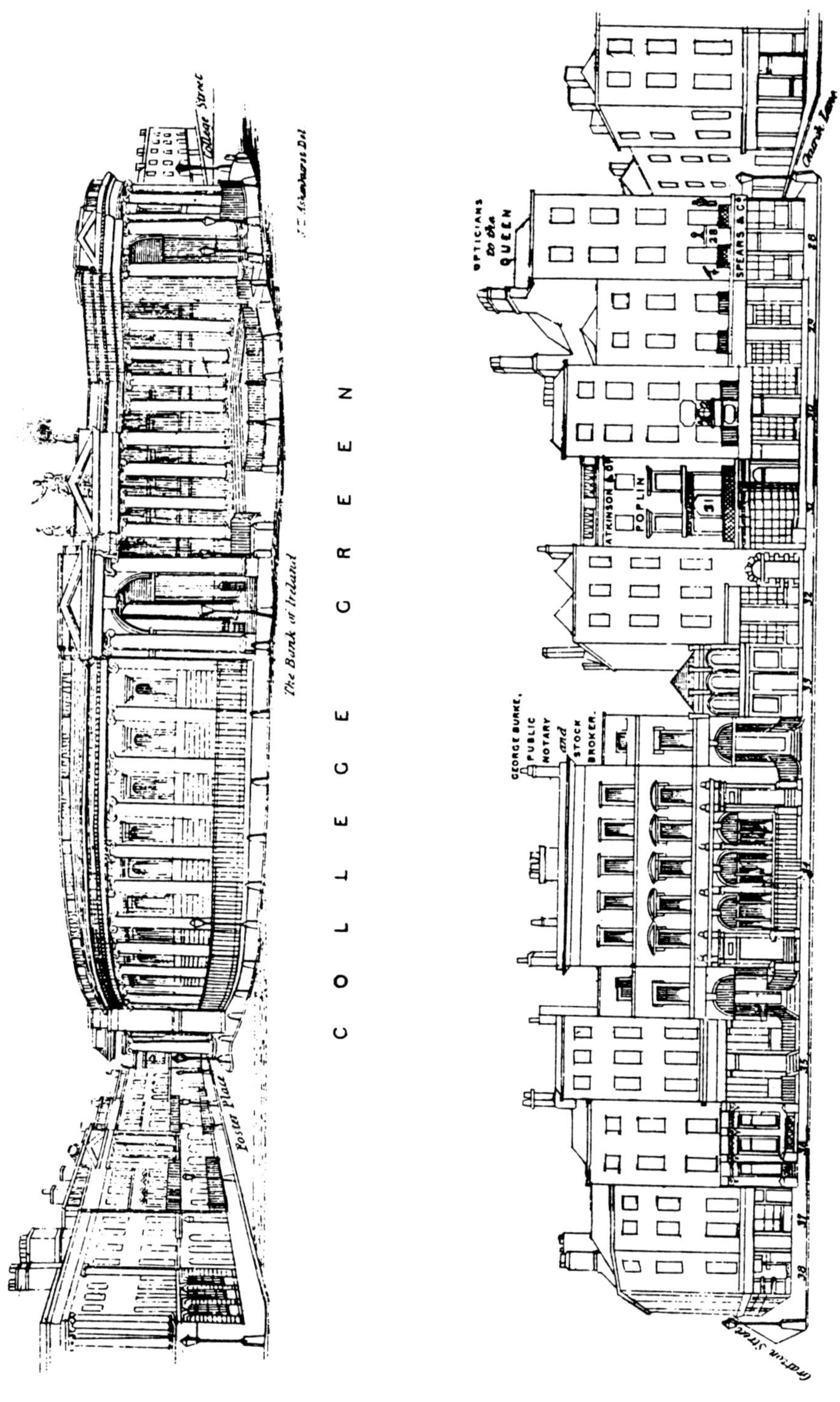

4.7. *Spears & Co.'s shop in College Green, as illustrated in Henry Shaw's* New City Pictorial Directory, *Dublin, 1850. National Library of Ireland.*

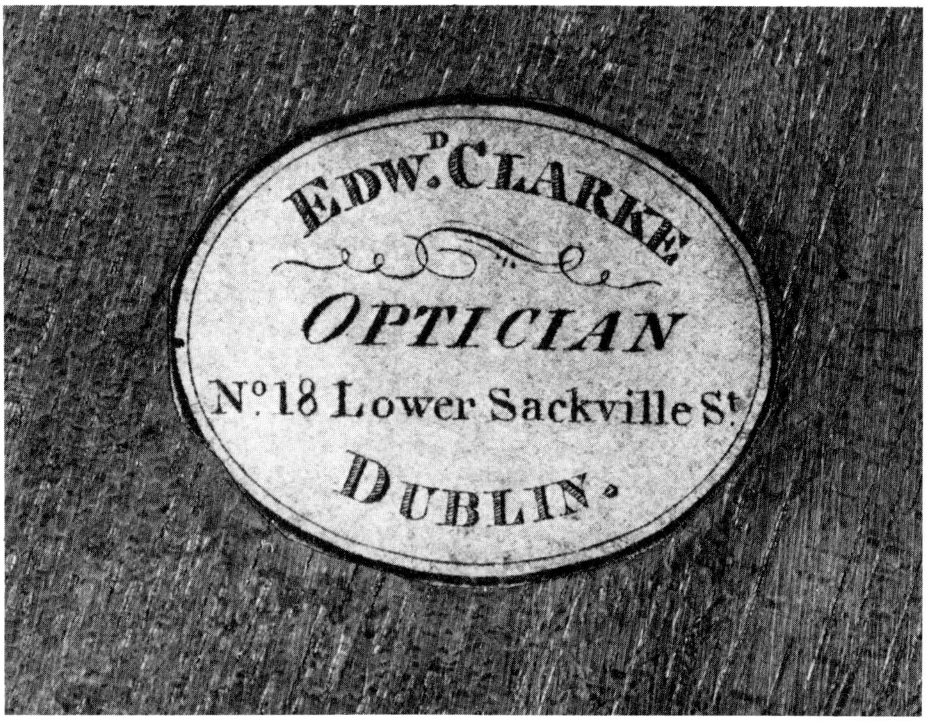

4.8. Trade card with microscope by Edward Clarke, Dublin, c.1810. Royal College of Surgeons of Ireland.

> PATENT PERISCOPE SPECTACLES
> RICHARD SPEAR, 23 Capel Street, Dublin,
> informs his Friends and the Public, that he has just received from London, a great variety of the NEW PATENT PERISCOPE SPECTACLES, which has gave [sic] such general Satisfaction to those who used them, both here and in England, and are found to preserve the Sight much better than any Spectacles before in use.
> N.B. Just received from London, a great variety of Adam's PATENT MILITARY TELESCOPES, and CARY'S NEW GLOBES.

Yet despite buying-in stock from other well-known London wholesalers, Spear was for a time in partnership with another maker whose inventiveness later became renowned. Between 1815 and 1817, at Spear's premises in College Green, he joined forces with Edward Clarke; after their two-year venture had split, Clarke, whose name was absent from the directories for a year or two, re-emerged as an optician in Lower Sackville Street in 1819 (*Fig.* 4.8), changed the name of the firm to E. Clarke & Co. at 83 Dame Street from 1823 until 1832, and then went to England. Of surviving instruments, a surveyor's compass signed "Speare & Clarke" has been noted[63]; two microscopes, two mechanical dials, and a sundial with Dublin addresses have been recorded for Edward Clarke[64].

It can be presumed that Clarke went to London to pursue more lucrative markets for his skills. Certainly, an instrument maker, Edward Marmaduke Clarke, emerges there, working with Watkins & Hill in 1833[65]. However, there was also an Edward S. Clarke, who contributed papers on electricity to the Royal Irish Academy in the early 1840s[66]. Because "Edward Clarke" of Dublin did not use his middle initial, it is difficult to be certain of his identity. However, he is likely to have been the former, for a number of reasons. His biography, as it appears in a late nineteenth century work, contains the following facts: he was born in Ireland, "took an energetic part in formation of Mechanics' Institution, Dublin 1826; occupied a prominent position at Science and Art exhibition in the King's Mews, London 1830; established London Electrical Society, May 1837 . . .; projected and promoted the [Royal] Panopticon [of Science and Art] in Leicester Sq. London . . . opened 16 March 1854, resident and managing director to 1856"[67]. However, this account does not say anything about a career as a scientific instrument maker, either in Dublin or London.

From various addresses in London, E.M. Clarke, "magnetician"[68] issued price lists of instruments which he manufactured: these show that he was particularly interested in electricity, and the name of the Irish inventor Nicholas Callan (1799-1864) is linked with specific pieces of apparatus[69]. Callan was ordained priest in 1823, and in 1826 was appointed Professor of Natural Philosophy at St Patrick's College, Maynooth, where he spent the rest of his life[70]. However, his discoveries in electromagnetism put him into the first rank of scientists: "he built large batteries and constructed huge electromagnets, invented the induction coil in 1836, and in 1838 discovered the principle of self-excitation in dynamo-electric machines"[71]. Credit for these usually goes to Ruhmkorff for the invention of the former in 1851, and Siemens for the latter in 1866. That there was a strong link between Clarke and Callan is shown by the presence at Maynooth of examples of the "Maynooth Battery", which bear the inscription "MAYNOOTH BATTERY E.M. CLARKE MAKER 428 STRAND LONDON" (*Fig*. 4.9)[72]. Callan's ideas reached a larger audience through *Annals of Electricity*, the journal conducted by William Sturgeon (1793-1850), a lecturer on natural philosophy who spent much of his time in London with the "magneticians" of the Strand[73].

4.9. "Maynooth battery", sold by E.M. Clarke of London, c.1845. St Patrick's College, Maynooth.

Before Clarke had been in partnership with him, Spear had been involved in at least one enterprise where he manufactured an instrument to a specified design: this was William Speer's hydrometer, launched in 1802[74]. The previous year Speer (who seems to have been no relation to his near-namesake), had been appointed to the office of Supervisor and Assayer of Spirits in the Port of Dublin. He was very unhappy with the current state of affairs, in which several different instruments were used to assess spirit values for Revenue purposes. He was not alone in this, for the Act of 1787 (Act 27 Geo. III, c. 31), although it legalised Clarke's hydrometer, called for further experiments to be carried out so that the most efficient and suitable instrument could be selected[75]. (This metal hydrometer had been invented about 1725 by John Clarke, a turner and engine maker, of Charing Cross, London; it is unlikely that he was related to Edward Marmaduke Clarke[76].) After three reports, a committee of enquiry was set up in London in 1802, and advertisements were placed in the press inviting the submission of instruments for examination. Ten makers responded. It is worth noting that a Mr Saunders, an instrument maker representing the Irish authorities, both sat on the committee and submitted an instrument for examination. He was probably Thomas Saunders, who in 1810, advertised[77]:

> HYDROMETERS AND SACCHAROMETERS
> THOMAS SAUNDERS, No 6 CHURCH-LANE, College Green,
> MATHEMATICAL INSTRUMENT MAKER to the Right Honorable and Honorable His Majesty's Commissioners of Inland Excise and Taxes in Ireland, begs leave to inform his Friends and the Public, that encouraged by the extensive patronage and universal approbation conferred on his HYDROMETERS, and stimulated by the success which has attended his labours, to reduce them to a standard of accuracy, he has made such arrangements in this branch of his business, as to enable him to meet an increasing demand, for which purpose he has fitted up a Workshop and Apparatus entirely for the manufacture of his improved HYDROMETERS and SACCHAROMETERS; and he has now ready for sale, a number of those instruments, together with others upon DICAS'S, RICHARDSON and CLARKE'S principle, which in point of workmanship, minute adjustment and extreme accuracy, equal to any ever made in England; also an extensive variety of Mathematical, Optical and Philosophical Instruments as usual, on the most moderate terms.

No examples of Saunders' hydrometer, nor his saccharometer, have so far been recorded although other instruments signed by him are known[78].

Neither Thomas Saunders' nor William Speer's instruments were chosen by the committee, who decided that the example submitted by Bartholomew Sikes should be that adopted: in fact this recommendation was not legalised until 1816, and needed a further Act of Parliament in 1818. This interval allowed competing instruments to be marketed, as in Saunders' advertisement above. Speer also invented a saccharometer, which was "made and sold by Richard Spear, Mathematical Instrument Maker, Capel-Street, Dublin, (By special Appointment to the Patentee). Where also Mr. Speer's Patent Hydrometer may be had"[79]. This was advertised in the press by Richard Spear[80]:

> TO THE BREWERS AND DISTILLERS, Of the United Kingdom of Great Britain and Ireland,
> RICHARD SPEAR
> Mathematical and Optical Instrument Maker, 23 CAPEL STREET, DUBLIN The only Person in Great Britain and Ireland authorised by the Patentee Mr WM. SPEER, to manufacture and vend the New Patent Saccharometer and Patent Hydrometer. Begs leave to inform the Brewers and Distillers of the United Kingdom, that he has ready for Sale the Patent Saccharometer and Hydrometer, invented by Mr Wm Speer which has received the Approbation of the Royal Irish Academy, the Dublin Society, and all of the scientific and

eminent Brewers and Distillers in Great Britain and Ireland. From the simplicity and accuracy of those Instruments they far exceed any Instrument of the kind ever before in Use, and the greatest Benefits must result from their Application.

Despite the large number of instruments signed by Richard Spear recorded, no examples of William Speer's instruments by him have so far come to light, although an example of his hydrometer, signed by William Harris of Holborn, London, is in the Science Museum, London[81]. Spear's pieces range from surveying instruments, including a waywiser, through optical goods to barometers; while still at 23 Capel Street, Spear had an appointment as "Instrument Maker to His Majesty's Ordnance" and a number of gunnery instruments are so inscribed[82].

Spencer, Robinson and photography

In the middle of the nineteenth century three further instrument makers of importance began work in Dublin: Thomas Grubb, John Spencer and James Robinson. Grubb, whose first known instrument was completed in 1835, is of such significance that he is discussed in greater detail in Chapter Six. John Spencer set up his business in Aungier Street about 1845. In 1864 he took his son into partnership; latterly they traded in Grafton Street and Nassau Street[83]. Spencer & Son were famous for their optical goods, and for their surveying levels. John H. Jellett (1817-1888), Professor of Natural Philosophy at Trinity College, invented a new optical "saccharometer" in 1863, because he felt that it should be "capable of giving more accurate results than those obtained by the instrument of Soleil"[84]. This used a new analysing prism, also developed by Jellett and described by him previously before the Royal Irish Academy; it is interesting to note that they were not made in Dublin[85]: "The prisms here described were

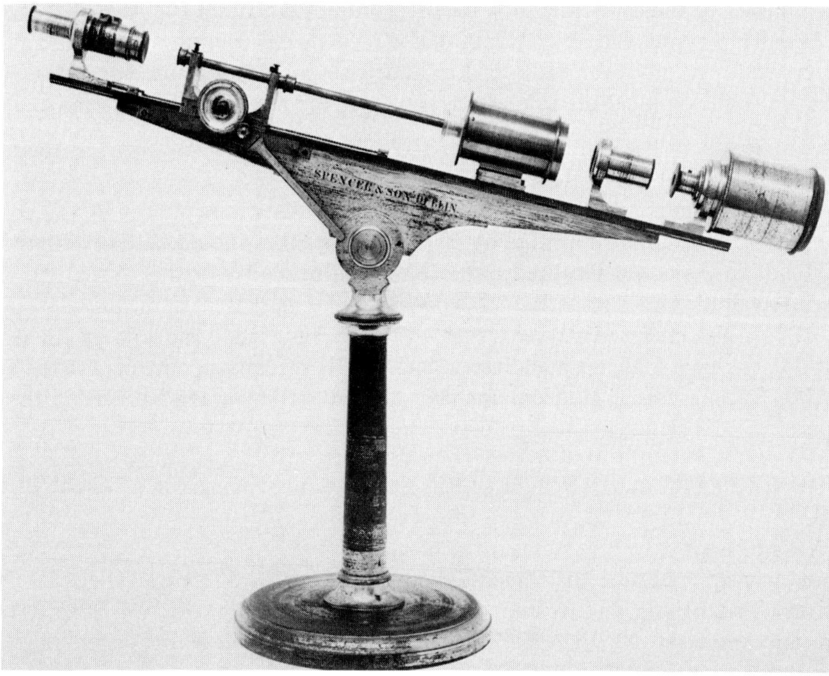

4.10. Jellett's "saccharometer" by Spencer & Son, Dublin, c.1865. Museum of the History of Science, Oxford.

executed by Mr J. Bryson of Princes Street, Edinburgh, in a manner which left nothing to be desired". Bryson had inherited a stock of Nicol prisms from their inventor William Nicol, and optical goods was one of his specialities[86]. However, by the time Jellett felt ready to develop his saccharimeter, he turned to a Dublin firm[87]: "the instrument here described was constructed by Messrs. Spencer and Son, of Aungier Street, to whose ability both in carrying out the instructions given to them and in suggesting methods for overcoming practical difficulties, the author is much indebted" (*Fig*. 4.10).

Spencer & Son exhibited Jellett's saccharimeter at the Dublin Industrial Exhibition of 1865, together with a cathetometer, for measuring vertical heights; a railway transit theodolite; levels and levelling staffs; air pumps; Cruise's endoscope (for viewing cavities inside the human body); and an equatorial stand[88]. The jury's report approved[89]:

> We would also direct especial attention to the originality of contrivance and refined accuracy of workmanship displayed by Messrs. J. Spencer & Son, in the collection of physical apparatus exhibited by them, particularly in a cathetometer, of rare excellence; in a Jellett's saccharometer [sic], of very beautiful design and execution; and in an air pump, simple in the arrangement of its parts, without valves between the pumps and the receiver, and free from regurgitation of air . . .

The firm was awarded a medal for the cathetometer, Jellett's saccharimeter and air pump, and honourable mention for their surveying levels. They exhibited Jellett's saccharimeter again in 1872, at the Dublin Exhibition of Arts, Industries and Manufactures, where the instrument was priced at £29[90], and again at the 1876 Loan Exhibition at South Kensington[91]: at least two examples are known in collections[92].

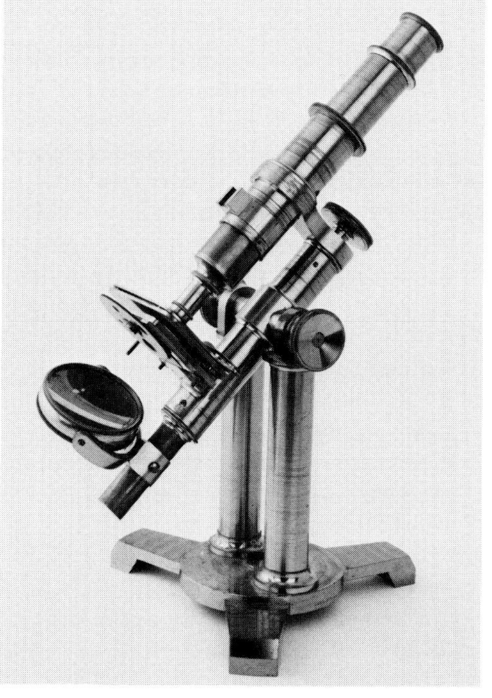

4.11. Compound microscope, unsigned, with hidden components scratched "Spencer Dublin", c.1860. Trustees of the National Museums of Scotland.

4.12.(a) Medical induction coil, signed "James Robinson Dublin". Clongowes Wood College.

4.12.(b) Detail of Robinson coil.

Spencer & Son also produced a number of microscopes (*Fig.* 4.11) and polarising microscopes[93]. Other sciences which had only used instruments to a limited extent were beginning to adopt them; for example, the Director of the Geological Survey of Ireland purchased petrological microscopes in 1871[94]. Businesses such as Spencer & Son were able to expand their range with that of local demand. Their heliostat, designed by George Johnstone Stoney in 1869, which they were offering for six guineas in 1872, similarly appears in a number of collections[95]. However, as with so many other firms, a major line appears to have been surveying equipment, for which they received a certificate of merit at the Cork Exhibition of 1883[96]: they displayed an improved dumpy level. Mason suggested that this was Spencer's own improvement, which was "copied and produced 30 years later by a big London firm who 'boomed' it as their own original design"[97]. Further proof has yet to be uncovered to substantiate or refute this accusation.

Even less is known about the firm of James Robinson which first appeared in 1845, and remained based in Grafton Street; the name changed to James Robinson & Sons in 1885, and the business continued well into the twentieth century. However, it would appear that they were more retailers than manufacturers, judging by their title "optician, polytechnic museum, photographic gallery, etc.", and this is born out by some of their later pieces such as microscopes bought in from E. Leitz of Wetzlar[98].

However, it seems that photography was one of their main lines of business, and James Robinson himself was involved personally in photographic circles from a fairly early date[99]. The history of photography in Ireland began as early as 20 September 1839 when Francis S. Beatty, an engraver living in Belfast, wrote to the *Belfast News-Letter* to claim success with the daguerreotype[100]. The first photographic studio opened in Dublin in October 1841 over the entrance to the Rotunda Rooms in Sackville Street, run by Le Chevalier Doussin Dubreuil: he was bankrupted by 1844. The longest surviving studio during the first ten years of photography was run by "Professor" Leon Glukman, who won a silver medal from the Royal Dublin Society

4.13. Dublin Photographic Society, committee for 1856-57. Thomas Grubb is seated in the front row, third from left; J. Robinson is standing in the back row, second right. Photographic Society of Ireland.

in 1847 for his daguerreotypes. Not much work by identified Irish calotypists survives; an exception was John Shaw Smith, who toured Europe and the Middle East in 1851-1852, and brought home about three hundred successful landscapes. A group of Dublin enthusiasts met, initially in the chemist's shop of William Allen in Henry Street, until the formation of the Dublin Photographic Society at a meeting in Leinster House (then home of the Royal Dublin Society) on 8 November 1854. Amongst the members were Thomas Grubb, Leon Glukman, James Robinson and John Shaw Smith (*Fig.* 4.13). The Society changed its name in 1858 to the Photographic Society of Ireland, but became dormant from 1860 until 1879; by 1894 interest had revived sufficiently for the annual Photographic Convention of the United Kingdom to be held in Dublin when the Presidential Address was given by Sir Howard Grubb, President of the Photographic Society of Ireland.

Robinson was sufficiently interested in photography to exhibit examples of his work in 1853; and at least one wet-plate camera sold by him is known[101]. It is probable that J.V. Robinson, who played such an active part in the Photographic Society of Ireland in the 1890s, and was in business as a "manufacturer of photographic apparatus" between 1885 and 1892 at 39 Lower Sackville Street, was a close relation of James Robinson[102].

Spencer, Robinson, Mason, Yeates; these are all English names, as are Stokes and King. The first large influx of English settlers had been in the reign of Elizabeth I, and Englishmen continued to migrate to Ireland thereafter. The guild structure had been used to limit the economic activity of the Catholic Irish, but this form of suppression had not been particularly successful. By the nineteenth century traders with Irish names are common in the Dublin street directories. An examination of contemporary politics enables us to shed a little more light in this area. After Daniel O'Connell achieved the emancipation of Catholics in 1829 he began agitation for the restoration of the high parliament in Dublin. In 1832, 1835 and 1837 there were closely contested elections for Dublin, and there was no secret ballot. From two published lists of voters it is possible to say a little about the political views of some Dublin instrument makers[103]. Several of them did not have the lowest qualification, that of being a £10 leaseholder. There are also difficulties in distinguishing between some of the voters: for example, two had the name James del Vecchio, and one voted for O'Connell, one against. James Lynch III, John Hicks, and Peter Mackey voted for O'Connell, as the Irish names of the first two might suggest. Thomas Mason, William H. Nelson and George Yeates were predictably against O'Connell[104]: it is highly likely that they were all Protestants.

It is perhaps worth emphasising that the migration of craftsmen was a two-way process; we have already mentioned that the Stokes and Mason families were originally of English extraction. We have also noted that members of the Yeates family moved to London to find work, as did Edward M. Clarke. In an advertisement of 1856, Elias Solomons, Optician and Spectacle maker of 27 Old Bond Street, Piccadilly, London, states that his improved telescopes can also be obtained at 19 Nassau Street, Dublin; which was the address of M.E. Solomons, optician, between 1856 and 1905. The relationship between the two Solomons has yet to be unravelled[105]. It is possible that once more detailed work has been done on instrument makers in provincial centres, further migration patterns will emerge; the Italians seem to have been particularly peripatetic right across Europe[106]. Irishmen also took their skills abroad: for example, Alexander Megary (1790-1850) was born in Ireland and appeared in the New York City directories as a mathematical instrument maker between 1827 and 1850[107]. James Joseph Hicks (1837-1916) was born at Ross Carbery in Co. Cork. It is not known if, or how, he was related to the other Irishmen surnamed Hicks who made and sold instruments in Dublin and elsewhere around Britain[108]. After leaving school, he was apprenticed to L.P. Casella in London, and later became his foreman. By about 1861 he was in business on his own account at 8 Hatton Garden, where during the nineteenth century his wholesaling firm became one of the largest in the British Isles[109]. A staunch Catholic, he presented meteorological instruments to the Vatican, and was made a Knight Commander of St Gregory[110]. Doubtless further research will uncover other migrating instrument makers.

International exhibitions

The second half of the century was littered with international exhibitions of greater or lesser splendour, imitating variously the Great Exhibition held in Hyde Park, London, in 1851. Only one Irish instrument maker exhibited there: George Yeates, who was described as "Proprietor and in part Designer and Manufacturer". He exhibited two barometers, two theodolites, an air pump, two prismatic compasses, an optic square and various spectacles[111]. The juries reported that he "exhibits a very portable small theodolite, with good divisions, well-made, and strongly put together . . . all these instruments are well-made, and deserving of Honourable Mention"[112]. Dublin was the first city to follow London's example, in 1853, and the exhibition proved something of a mixed blessing[113]: it made a large loss. George Yeates, James Robinson, Richard Spear, W. Nelson and Thomas Grubb all exhibited a large array of instruments[114]. At Paris in 1855 Grubb and Yeates were both represented, and the Earl of Rosse sent a model of his great telescope[115]. The London exhibition of 1862 did not have the panache or public appeal of the Great Exhibition. Grubb was represented by a 12-inch achromatic telescope, and Yeates by a "public barometer" and thermometer, "for the use of agricultural districts and fishing towns"[116]. One had been installed at Carrickmacross, and others in fishing towns on the estates of the Marquess of Lansdowne[117]. There were no Irish exhibitors at Paris five years later[118]. The telescope which Grubb displayed at Paris in 1878 is discussed in Chapter Six. The Society of Arts sent a deputation of artisans to report on the Paris Universal Exhibition of 1878, and Michael Lambert was sent with the Dublin group to assess the optical instruments[119]. Some of his remarks are very revealing about the conditions of the trade in Dublin compared with those in France and England (Germany did not exhibit)[120]:

> The social system adopted in the French workshops differs very much from ours. In the first place the men work longer hours (11 hours each day) and they take but one meal during that time . . . In the next place they do not keep so closely at work as our men do, and they are not prevented from smoking or chatting on any subject they please during working hours . . . I do not think that they get as much work done in the eleven hours as we do in nine. This, I conceive, is in a great measure to be accounted for by their not having a recognised day of rest . . . They cannot, therefore, be expected to work as diligently as men who rest regularly once a week, and take their recreation at stated times. There is one great advantage which the artisans in Paris enjoy that is denied to our men, that is facilities for educating themselves. At every side they have museums, and exhibitions of the arts and manufactures of every nation on the earth from the earliest ages to the present day, and libraries in which there are to be found works on every subject. To any of these the artisan can repair during the day or in the evening, and see the instruments used in every science, and the machinery, or models of machinery, employed in every art and manufacture, from their first development to their latest improvements; and he can afterwards enter a library, and read their history or study their theory, according to his inclination.
> With such opportunities open to them can we wonder that the French workmen have acquired for themselves so high a reputation for intelligence and skill?
>
> In Dublin, the artisan after his day's work is done finds no place where he can spend an hour in intellectual recreation. He has no public library, the museums are, with one exception, private property, and entrance to them can only be obtained through a letter of introduction . . .

Paris appears to be the last international exhibition bar one which Grubb attended. Having missed major shows in Paris in 1889 – for which the Eiffel Tower was built – and 1900, Grubb sent a telescope to the St Louis Exhibition of 1904. This event was held to celebrate the centenary of the Louisiana Purchase. Howard Grubb took it as an opportunity to exploit the expanding market in the United States for large observatory telescopes, domes, and other astronomical apparatus. He won a Grand Prize[121].

There was another Dublin exhibition in 1872. Only three local instrument makers were represented. Yeates contributed an astronomical telescope and some philosophical instruments, and Grubb showed models of his large telescopes. Spencer & Son exhibited a heliostat and a Jellett saccharimeter[122]. The presence of a tradesman at an exhibition implies that he made his own goods – nationalism and local loyalty strongly discouraged the display of "bought in" goods unless they were displayed separately as imports – and the absence of makers other than the three already named indicates that they were probably the only three manufacturers in Dublin at this time.

In 1885 the International Inventions Exhibition was held in South Kensington. Its purpose was to display progress in the 25 years since the exhibition of 1862. There was a very large number of exhibitors, and the catalogue gives a good listing of the names of the inventors represented rather than a clear idea of what, exactly, they displayed. The two Dublin instrument makers who sent their products were Yeates & Son and F.M. Moore[123]. Yeates showed a standard metre, a standard yard, a patent standard mercurial pressure gauge, an electrical rain gauge, and an electrical anemometer. Moore's contribution was an improved mariner's compass, but the "improvements" proved to be an infringement of Sir William Thomson's patents No. 1339 and 1339* of 1876; Moore admitted this, but transgressed again in 1888. The following year, Thomson successfully prosecuted Moore[124].

4.14. Detail of the Dublin Exhibition of 1865, showing the Yeates display. Trustees of the Science Museum, London.

Applied science in Ireland

Three Dublin scientists provided instruments for the Inventions Exhibition of 1885. James Dillon of 38 Dawson Street, a civil engineer, sent hydrographic surveying apparatus[125], which he patented[126]. Walter Hartley, Professor of Chemistry at the Royal College of Science for Ireland, showed an ultraviolet spectrograph and a portable direct vision spectrometer[127]. Hartley, born, like that distinguished amateur chemist, Samuel Johnson, in Lichfield, worked in Dublin from 1879 to 1911, and made many important advances in spectroscopic technique[128]. One of his ultraviolet spectrographs, made in part by Yeates, is in the Science Museum, London[129].

John Joly (1857-1933) was the third scientist who displayed his inventions at London in 1885 – apparatus for reading barometers and thermometers at a distance[130]. This apparatus was an application of telegraphic apparatus to pure science[131]. When a barometer reading, say, was required, a length of wire was lowered from a known point within the upper part of the barometer tube, and at a known rate: on making contact with the surface of the mercury an electrical circuit was completed. The time taken for the wire's descent was measured by counting the number of make-and-breaks performed by a rheotome situated beside the observer. From the measured time the barometric reading could be calculated. Joly acknowledged the work of Yeates in preparing the instrument. Although a most distinguished geologist, Joly invented a wide selection of scientific instruments. These included a specific gravity balance[132], a photometer[133], and the meldometer, for measuring the melting point of minerals[134]. The last of these was made by Yeates.

There was a notable interest in the apparatus of science, and in applied science and industry in general, among Irish scientists in the second half of the nineteenth century. George Francis Fitzgerald, chiefly known as a theoretical physicist, wrote also on the design of dynamos[135], and had a conspicuous though unsuccessful career as a pioneer flyer: he is remembered as "Flightless Fitzgerald". Almost equally distinguished as a theoretician, George Johnstone Stoney invented a new musical instrument[136]. John Joly was another example, as was G.G. Stokes, who gave extensive assistance to the Grubbs (see Chapter Six). Anthony Trail, later Provost of Trinity, with his brother built the Giant's Causeway Tramway in 1883 – the first electric railway in Great Britain and Ireland[137]. It drew power from a very early hydro-electric power station on the River Bush. Similarly, geologists working in Ireland were particularly interested in economic geology. One of them wrote, "I am strongly of the opinion that we should never lose an opportunity of bringing before the scientific public any natural product of this country which is likely to be of commercial value"[138].

The intellectual source for this concern with the resources and the technology for exploiting them was Sir Robert Kane's *The Industrial Resources of Ireland* (1845). One of Kane's ideas was the Museum of Irish Industry, founded in 1846. The Museum had a teaching role, which in 1867 was passed, on its foundation, to the Royal College of Science for Ireland[139]. Its aim was to teach applied science: it was felt that pure science was already adequately covered by Trinity[140]. "The Royal College of Science supplies, as far as practicable, a complete course of instruction in Science applicable to the Industrial Arts, especially those which may be claimed broadly under the heads of Mining, Engineering and Manufacturers . . ."[141]. Its professors were pioneers of industrial research. The name of W.N. Hartley has already been mentioned; W.F. Barrett was his contemporary as Professor of Physics. Barrett devised various instruments, including a colorimeter which he had made by his countryman, J.J. Hicks, instrument maker in London[142]. In 1882 Robert Hadfield of Sheffield produced the first sample of manganese steel. It proved to be far harder than ordinary steel. Needing a full understanding of the physical properties of manganese steel, he turned to James Bottomley of Belfast, nephew of Lord Kelvin. Bottomly sent a specimen of the steel to Barrett, who made extensive studies of it[143], and subsequently he investigated other special steel with Hadfield. Barrett had some of his experiments repeated at Inchicore Works by H.G. Ivatt of the Great Southern and Western Railway.

The manufacture of instruments in Dublin

By the middle of the nineteenth century fewer and fewer instrument "makers" actually had the skill to create an instrument from scratch. Conversely, wholesale manufacturers were becoming increasingly important. One wholesaler, James Parkes of Birmingham, explained in the introduction to a catalogue which he issued in 1867:

> The improvements made in modern machinery, (especially in the Midland Counties of England), and the facilities which are afforded by extensive railway intercommunication, have done much to break down . . . local peculiarities. Manufacturers, therefore, instead of depending on a mere name, or old reputation, have now to study how they may skilfully *combine* the special advantages peculiar to some localities, with those of others; so that, by the aid of machinery improvements, they may obtain a superior total result. This, we submit, has been our aim . . . Skilled London workmen are also employed by us on certain parts of Instruments used for Optical and Mathematical purposes; so that, with the additional aid of our special machinery improvements, we have been able to produce a really superior article at a far more moderate cost[144].

Later, the same firm boasted that they were able to reduce their prices because they had proceeded with mechanising the manufacturing process, and had introduced the division of labour[145]. At the end of the century one of the largest producers of microscopes in London announced that he was importing plant from Germany and America[146]. The increase in capital required to continue manufacturing instruments reduced many firms to the status of mere vendors.

Spencer & Son also retailed instruments which they almost certainly did not make, such as a pair of early clinical thermometers (c.1870), according to the design of Sir William Aitken. However, the label on their case indicates that in addition to the shop in Grafton Street, Spencer & Son also had a "Manufactory" at 13 Aungier Street[147].

In assessing the supply of laboratory instruments in Ireland towards the end of the nineteenth century we have a valuable guide in the archives of the Royal College of Science for Ireland. The book of "Lecture Outfit Orders" from 1868-1879[148] shows that more than half of the money spent on instruments and apparatus went to James Rigg of Chester, William Ladd of London, and for chemical apparatus to four London firms: J.J. Griffin, A. & M. Zimmerman, M. Jackson and Townson & Mercer. One particularly interesting entry in 1876 is for £30 for an induction coil – it must have been a large one, since Yeates & Son priced their largest ("length of spark 5 in") at £20[149]. But the coil was not bought from Yeates & Son, but from Horatio Yeates, who last appears in the Dublin street directories for 1864, after which he was based in London.

The Royal College of Science for Ireland's accounts for 1883-1885 also survive[150]. They contain the following payments for instruments:

Yeates & Son, Dublin	£25	11	3
J. Robinson, Dublin	5	6	11
T. Mason, Dublin	2	12	6
F.M. Moore, Dublin	5	0	0
J. McVey [Dublin]	6	5	8
R. Field, Birmingham	1	18	0
W. Hume, Edinburgh	4	19	11
A. Apps, London	1	1	0
J.J. Hicks, London	2	6	6
Baird & Tatlock, London	2	8	0
C. Zeiss, London	1	1	0
Total	£58	10	9

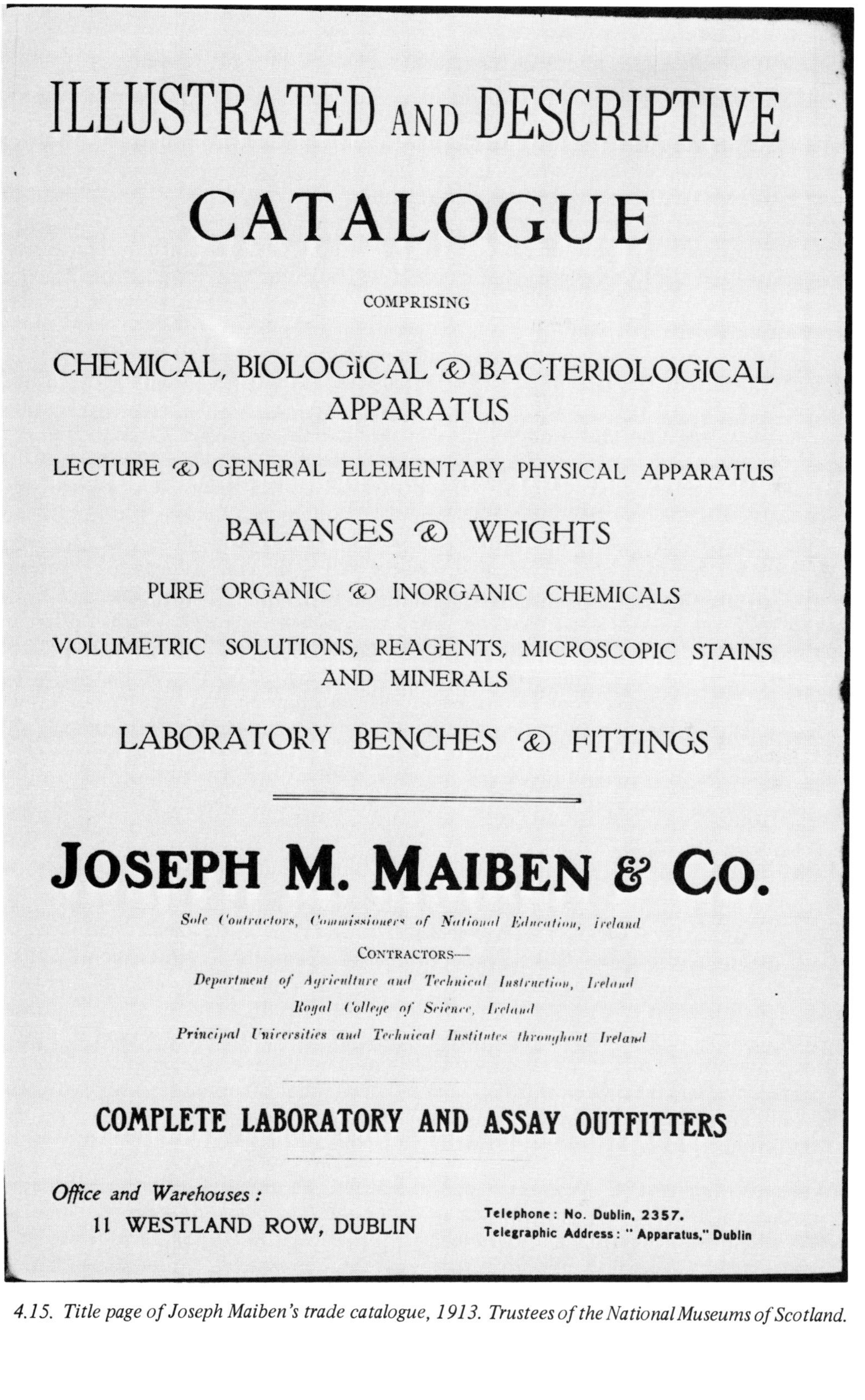

ILLUSTRATED AND DESCRIPTIVE

CATALOGUE

COMPRISING

CHEMICAL, BIOLOGICAL & BACTERIOLOGICAL APPARATUS

LECTURE & GENERAL ELEMENTARY PHYSICAL APPARATUS

BALANCES & WEIGHTS

PURE ORGANIC & INORGANIC CHEMICALS

VOLUMETRIC SOLUTIONS, REAGENTS, MICROSCOPIC STAINS AND MINERALS

LABORATORY BENCHES & FITTINGS

JOSEPH M. MAIBEN & CO.

Sole Contractors, Commissioners of National Education, Ireland

CONTRACTORS—

Department of Agriculture and Technical Instruction, Ireland

Royal College of Science, Ireland

Principal Universities and Technical Institutes throughout Ireland

COMPLETE LABORATORY AND ASSAY OUTFITTERS

Office and Warehouses :

11 WESTLAND ROW, DUBLIN

Telephone: No. Dublin, 2357.

Telegraphic Address: "Apparatus," Dublin

4.15. Title page of Joseph Maiben's trade catalogue, 1913. Trustees of the National Museums of Scotland.

The largest single orders were to Yeates & Son for a specimen of Iceland spar, and to Moore for a reflecting circle. Assuming McVey was in Dublin, the Royal College of Science for Ireland thus spent 76% of their expenditure on instruments in Dublin during this period.

In considering instrument making for teaching or research, one further figure must be considered – the laboratory technician, or "mechanic". James Lynch II had filled this role for the Natural Philosophy Department of Trinity earlier in the century. Unfortunately laboratory technicians are usually inconspicuous, but one other Irish example can be given – Henry Pye. William Pye (1847-1921) was a highly skilled craftsman who from 1880-1898 was foreman of the Cambridge Scientific Instrument Company. William and his son Henry left the company in unusual circumstances. Another son, W.G. Pye, had set up a rival instrument making concern in Cambridge, and the firm believed that it would be unwise to have members of the family working for them. Henry Pye moved in 1899 to work for W.F. Barrett at the Royal College of Science for Ireland[151].

The last firm to enter the scientific instrument market before the First World War was that of Joseph M. Maiben. The substantial size of his business can be judged from the catalogue he issued in 1913 (*Fig.* 4.15)[152]. Nowhere, however, does Maiben suggest that he was anything more than a supplier of instruments. Study of the illustrations in his catalogue reveals the following instrument makers were represented: John Browning of London, Schmidt & Haensch of Berlin, R. Winkel of Göttingen, Ernst Leitz of Wetzlar, Carl Reichert of Vienna, Carl Zeiss of Jena, J.J. Hicks of London and the Cambridge Scientific Instrument Company[153]. The balances carry the names of Salter, Sartorius, Bunge and Oertling[154]. The slide rules are clearly marked "A.W. Faber made in Bavaria"[155]. There are Leclanché cells by the India Rubber Co., Silvertown, London[156], and furnaces and many other pieces of apparatus by Fletcher, Russell & Co. of Warrington[157], and even an ore crusher by F.W. Braun of Los Angeles[158]. Maiben offered to supply on request a catalogue of microscope manufacturers[159]. It is safe to say that most instrument "makers" at this time – in, for instance, Scotland and the English provinces, apart from Birmingham – were suppliers too. The nature of instrument production had changed so much that by the end of the century only specialists could "make" instruments, and then only a limited range.

The decline of Irish science

It has been observed by several historians that Irish science entered a serious decline at the end of the nineteenth century[160]. Irish medicine, too, enjoyed a golden age in the middle of the nineteenth century[161]. When one sees the presidential address to the Royal Geological Society of Ireland in 1885, the reader inevitably considers whether the owl of Minerva was indeed taking flight in the dusk. Valentine Ball began by outlining a scheme for a series of monographs on mineral resources in Ireland, and then diverted himself into "A geologist's contribution to the history of ancient India" – at great length[162]. One of the channels by which science in Ireland was given financial support was through a grant passed annually from the government, through the Royal Society of London, to the Royal Irish Academy. It was restricted to pure science – a singular irony when there was a widely-perceived need for contributions from applied science to the solution of Ireland's economic difficulties. When one geologist was refused a grant by the Royal Irish Academy, he wrote a memorably intemperate letter[163]:

> I regret to find that the Committee look at Science in a different light to what they do in America, as they there consider *Science to be the Handmaid of Commerce*. May I ask how in America they could have learned the value of Asbestos, or where it was to be procured, if Science had not paved the way for Commerce? . . . I presume that the Science Grant was intended for the benefit of the country at large; and I would submit

> that no greater benefit could be conferred on the country than by spending money in the fostering and developing of a new source of industry.

The science grant tended to be given to naturalists whose work could be of little or no economic benefit to Ireland, and it fostered the highly competent group of biologists who flourished in Ireland at the end of the nineteenth century, and who have been admirably described by Robert Lloyd Praeger[164].

The decline of Irish science is so important that it must be discussed in some detail. Herries Davies has studied the numbers of papers being published on the earth sciences in Irish periodicals, and demonstrated that there was a very rapid decrease from about 1880 to the First World War[165]. Elsewhere, Herries Davies has suggested six possible explanations for the decline[166]. Having pointed out that a disproportionately high percentage of Irish scientists were Protestants, he draws attention to the falling Protestant population. In the 26 counties which broke away from the United Kingdom in 1922, the Protestant population had been 468,000 in 1861; it was to be only 144,500 a hundred years later. The part of the population which historically produced scientists was declining swiftly. It was a contemporary view that the ablest were leaving. One prominent Catholic, Robert Kane, addressed the inaugural meeting of the Statistical Society of Ireland in 1866 and, pointing out that the population of Ireland had fallen by 50,000 in the previous year, commented that Ireland had suffered "by the loss of that number of industrious and energetic members of the labouring classes, leaving behind them, a burden on the land, the idle, the imbecile, and the diseased"[167]. In an era when education was becoming widely available, there can be no doubt that scientific talent was being lost. The move of intellectual leaders to London was noticed by other commentators – including Walter Hartley's Dublin-born wife[168]. The second point is that this decline was closely related to an enormous emigration which had begun in the Famine; and that, although it is difficult to prove, it is likely that among those who left Ireland were many with initiative and a good education.

Thirdly, the standards of science in other countries were rising as science became a more competitive activity. When so many of the best were leaving Ireland, and remaining scientists found most of their time consumed by teaching, Ireland was in danger of becoming a scientific backwater. One chemist used a different metaphor to describe the difficulty. He praised the competitive system in Germany, and added, "from a multitude of such enthusiasts the great professors come. Great mountains are only found in mountainous countries"[169]. Fourthly, there was little financial aid available for science, especially since there was less industry in Ireland than in England or Scotland: this must have had an effect on the purchase of scientific instruments. The fifth point, that the disturbed state of the country must have discouraged the field scientists, has little bearing on the market for instruments – except that, as Herries Davies observes, fear of sabotage or armed insurrection caused the British government to move Grubb's works from Rathmines to St Albans. Finally, Herries Davies draws attention to the fact that the Irish government after 1922 was strongly nationalist, both politically and culturally, and that science must have suffered as attention was diverted to the study of Gaelic civilisation. It must be likely that the cultural shift had begun some 30 or 40 years earlier, when W.B. Yeats, Lady Gregory and Douglas Hyde began the Gaelic revival. It is difficult to know how to relate important events such as the founding of the Gaelic Athletic Association (1884) and the Gaelic League (1893) to the history of science: perhaps the conclusion we should draw is that the scientific and nationalistic cultures were entirely separate from one another, and as far as the inner workings of each of them were concerned, the other did not exist. Only when the new state had acquired self-confidence could science again flourish. It may be significant that the only Irishman so far to win the Nobel Prize, E.T.S. Walton, conducted his pioneer atom-smashing in Cambridge (1932), but subsequently returned to be Professor of Natural and Experimental Philosophy at Trinity.

References and Notes for Chapter Four

1. Thomas H. Mason, Dublin Opticians and Instruments Makers, *Dublin Historical Record*, **6**, 1944, 138; [Thomas H. Mason & Sons Ltd], *The Mason family business: a brief history 1780-1980*, [Dublin, 1980].
2. Mason, *op. cit.*, 140 (ref. 1).
3. E.G.R. Taylor, *Mathematical Practitioners of Hanoverian England*, Cambridge, 1966, 268, gives Margas as an apprentice of Nathaniel Adams; his original business in Rose Street, Westminster, failed in 1758 and he was listed as bankrupt in the *Gentleman's Magazine*, **28**, 1758, 197. A typescript by T.H. Court in the Science Museum, London, gives (on p. 207) the information that John Margas was in the Spectaclemakers' Company, and that he took Nathaniel Rawlins as apprentice on 3 October 1751. T.H. Court & M. von Rohr, Contributions to the history of the Worshipful Company of Spectaclemakers, *Transactions of the Optical Society*, **31**, 1920-30, 81, item 137 give him as "Marges", mentioning his bankruptcy as 4 April 1758. We are grateful to Dr John Chaldicott for this information. The Court Papers in the Science Museum Library, London, have some notes presumably made from the guild records: John Margas was the son of Jacob Margas, goldsmith; he was made free on 30 December 1742, but there was a problem; on 29 December 1743 it was reported that Margas had been working illegally for J. Cox, who had not paid the Company's dues. Besides Nathaniel Rawlins, Margas also took on James Jameson as an apprentice on 4 October 1753; he was made free on 1 January 1761. Court Notebooks, X/2, 118. Margas also took on as apprentices Edmund Say on 4 August 1743, and William Coltenridge on 4 October 1753 at the same time as James Jameson. Our thanks to Project Simon for this information. A stick barometer signed "J. Margas from London now in Capel-Street DUBLIN" has been recorded. We are grateful to Dr Charles Mollan for this information.
4. *Dublin Chronicle*, 10 May 1787, illustrated in [Mason], *op. cit.*, 3 (ref. 1).
5. Item 802 in *Exhibition illustrating some Irish contributions to the progress of the arts and science*, St Patrick's College, Maynooth, 1953; and item 801.3 in *Museum third Tostal display souvenir catalogue*, St Patrick's College, Maynooth, 1955.
6. [Mason], *op. cit.*, 4 (ref. 1). An octant, signed "Crichton London sold by Mason Limerick" offered for sale by Sotheby's 20 February 1985 Lot 74, indicates that Jonathan did not make all that he sold.
7. An example is in the National Maritime Museum, Greenwich, inventory no. C.153; another, incomplete, is in the Department of Engineering, Trinity College, Dublin.
8. James Apjohn, Formula for inferring the dew-point from the indications of the wet-bulb hygrometer, *London and Edinburgh Philosophical Magazine*, 3rd series, **6**, 1835, 182-188; On a formula for ascertaining the dew-point by the wet-bulb hygrometer, *London and Edinburgh Philosophical Magazine*, 3rd series, **7**, 1835, 266-274, 470-473; On the theory of the moist-bulb hygrometer, *Transactions of the Royal Irish Academy*, **17**, 1837, 275-282, 283-292.
9. W.E.K. Middleton, *The invention of the meteorological instruments*, Maryland, 1969, 126.
10. [Mason & Son], *op. cit.*, 19-22 (ref. 1).
11. Mason, *op. cit.*, 143-144 (ref. 1).
12. *Appendix to the Thirtieth Report of the Deputy Keeper of Public Record and Keeper of the State Papers in Ireland: an index to the Act or Grant Books and Original Wills of the Diocese of Dublin from 1800 to 1858*, Dublin, 1899, 724; also recorded in *Index to births, marriages and deaths in a file of the Hibernian Chronicle October 1769-April 1802*. Typescript by the Society of Genealogists, 1936.
13. Family trees of numerous branches of the Yeates and Dixons have been generously provided by Professor Kendal Dixon, grandson of Rebecca Dixon, née Yeates, from records kept by her son James Dixon. We are grateful to Dr Charles Mollan for providing us with copies of these.
14. Obituary, *Monthly Notices of the Royal Astronomical Society*, **37**, 1877, 159-160.
15. Andrew Yeates, On the notches, Y's or bearings for the pivots of transit instruments, *Monthly Notices of the Royal Astronomical Society*, **25**, 1865, 214-215.
16. Frederic Boase, *Modern English biography*, 3, London, 1965, 1560-1561.
17. The prismatic compass is in the Science Museum, London, inventory no. 1948-27; as is the Gunter's rule, inventory no. 1910-164. The transit theodolite was offered for sale by Christie's South Kensington 21 October 1982 Lot 94.
18. Although the family tree gives her name as "Elizabeth", a letter in the Court Papers states categorically that "Andw. Yeates married Caroline, Cornelius's daughter"; Court Papers, *loc. cit.*, (ref. 3), Y/4/2.
19. *Catalogue of the special loan collection of*

scientific apparatus at the South Kensington Museum, 3rd edn., London, 1877, 311, item 1304aa.

20. A theodolite is in the collection in the Department of Engineering, Trinity College, Dublin; a Ross-type monocular microscope was offered for sale by Sotheby's 10 May 1979 Lot 13.
21. Our thanks to Julian Holland for this information.
22. Much of this information has been used in A.D. Morrison-Low, Yeates & Son opticians and mathematical instrument makers to the University, in C. Mollan *et al.* (eds), *More people and places in Irish science and technology*, Dublin, forthcoming.
23. Trinity College Dublin, MUN/P/4/267/40.
24. Trinity College Dublin, MUN/P/4/214/71.
25. Trinity College Dublin, MUN/P/4/121/12 and 267/29.
26. In a private collection.
27. Discovered in an elbow polariscope in the Royal Dublin Society by Dr Charles Mollan. It advertises "Smith & Beck's Microscopes" – this firm traded under this name from 1847 to 1857.
28. *Ibid.*
29. Yeates & Son, *Illustrated catalogue of drawing, surveying and general engineering instruments*, Dublin, 1887, 22-23. The original of this, and the other Yeates catalogues referred to here, are in the Department of Physical Sciences, Science Museum, London.
30. Yeates & Son, *Illustrated catalogue of optical instruments and apparatus*, Dublin, 1880, 23. Italics in the original.
31. Yeates & Son, *Illustrated catalogue of electrical apparatus and appliances*, Dublin, 1877, 35.
32. Yeates & Son, *op. cit.*, 37-38 (ref. 30).
33. Yeates & Son, *op. cit.*, 29, 32-34 (ref. 29).
34. H.C. Casserley, *An outline of Irish railway history*, Newton Abbott, 1974, 20.
35. Ronald C. Cox, (ed.), *The School of Engineering, Trinity College Dublin: a record of graduates*, 5th edn., 1981, 1.
36. Richard Griffith, [Presidential Address], *Transactions of the Institution of Civil Engineers of Ireland,* **6**, 1859-61, 201.
37. Cox, *op. cit.*, 5 (ref. 35).
38. George Yeates, On an improved construction of theodolite, *Transactions of the Institution of Civil Engineers of Ireland,* **1**, 1844-45, 174-176.
39. George Yeates, A simple instrument for measuring vertical angles, *ibid.*, **2**, 1846-47, 53.
40. George Yeates, An instrument extending the utility of the dumpy level, *ibid.*, **2**, 1846-47, 70.
41. George Yeates, The optic square, *ibid.*, **2**, 1847-47, 126.
42. George Yeates, A new drainage instrument, combining the advantage of the theodolite and level, *ibid.*, **3**, 1847-49, 31.
43. Yeates & Son, *Apparatus for experiments in heat*, Dublin, 1883, 14, 17.
44. Yeates & Son, *op. cit.*, 16 (ref. 30).
45. *Ibid.*, 52, 55.
46. *Ibid.*, 28, 34.
47. Yeates & Son, *op. cit.*, 53-55 (ref. 29).
48. List of apparatus in the Natural Philosophy Department of the Royal Dublin Society. MS in the Royal Dublin Society. We are grateful to Dr Charles Mollan for drawing this to our attention.
49. An atmospheric railway based on Brunel's ideas ran between Kingstown (now Dunlaoghaire) and Dalkey, from 1844 to 1854 (Casserley, *op. cit.*, 28 (ref. 34)). The model (now lost) can be confidently dated to within those years.
50. Mason, *op. cit.*, 148-149 (ref. 1).
51. George Yeates, Meterological journal, from 1st January to 31st December 1843, *Proceedings of the Royal Irish Academy*, **2**, 1846, appendix, and in each succeeding volume.
52. George Yeates, On a meteor seen near Dublin, *Proceedings of the Royal Irish Academy*, **4**, 1850, 37-38.
53. Stephen M. Yeates, On a new table and formula for determining altitudes with the baometer, *Journal of the Royal Dublin Society*, **4**, 1866, 444-445. S.M. Yeates also took out a patent, no. 9244 of 1886, for improvements in surveyors' levels, besides others connected with tramways.
54. *Minutes of the Proceedings of the Dublin Microscopical Club*, **1**, 1866, i.
55. *Ibid.*, **1**, 1866, 66.
56. *Ibid.*, **1**, 1866, 55; **1**, 1868, 202.
57. S.P. Thompson, *Phillip Reis: inventor of the telephone*, London, 1883, 47.
58. B.M. Phadraig, Dublin one hundred years ago, *Dublin Historical Review*, **23**, 1969, 59.
59. A.M. Fraser, Statues on public buildings in Dublin, *ibid.*, **2**, 1939-40, 36.
60. C. Dickens, *Dombey and Son*, Penguin edition, London, 1974, 88.
61. S.A. Bedini, *Thinkers and Tinkers*, New York, 1975, 220-224.
62. *Hibernian Journal*, 22 July 1805, and intermittently for the rest of the year.
63. At St Patrick's College, Maynooth. Item 801.4 in *Museum third Tostal display . . ., op. cit.*, (ref. 5).
64. A Jones Most Improved microscope and a portable dial are in a private collection; a brass Culpeper type microscope is in the Royal College of Surgeons of Ireland, and a universal equi-

noctial dial was offered for sale at Sotheby's 26 July 1965 Lot 115. A brass garden dial signed "EDW'D CLARKE/10 Lower Sackville St./DUBLIN" was offered for sale by Historical Technology *Catalog* 106, Late Spring 1973, item 78.

65. Taylor, *op. cit.*, 359 (ref. 3). Taylor calls him Edward Montague Clarke.
66. Edward S. Clarke, On atmospheric electricity, *Proceedings of the Royal Irish Academy*, **1**, 1841, 166-170; On an improved form of the voltaic sustaining battery, and on the size proper to be given to the zinc element of sustaining batteries in general, *Sturgeon's Annals of Electricity*, **3**, 1838-39, 85-92; Improvements in the voltaic battery, *ibid.*, 314-315.
67. Frederick Boase, *Modern English biography*, **4**, London, 1965, 674-675. The 1830 exhibition was that of the Society of Arts: our thanks to Brian Gee for this information.
68. For instance, Description of E.M. Clarke's Magnetic Electrical Machine, *London and Edinburgh Philosophical Magazine*, 3rd series, **9**, 1836, is signed "Edward M. Clarke, Magnetician". Brian Gee (personal communication) points out that he used this title only whilst making and selling magnetico-electric machines, between about 1836 and 1840; after that, he styled himself "optician". Clarke was at 39 Charles Street, Westminster in January 1835; at 9 Agar Street, Strand in April and September 1836; at 11 Lowther Arcade, Strand from February 1837 until at least August 1839; at 428 Strand between February 1840 and May 1850. Among the advertisements at the end of *The illustrated handbook of the Royal Panopticon of Science and Art*, London, 1854, is one by Horne, Thornthwaite & Wood, 121 & 123 Newgate Street, stating that "E.M. Clarke Esq., having the entire direction of the Royal Panopticon of Science and Art has transferred the whole of his late Optical and Philosophical Business at 428 Strand to Messrs. H.T. and W. . . . who continue to supply all that class of Instruments for which E.M. Clarke was so noted". Our thanks to Dr John Chaldecott for this information. Brian Gee kindly tells us that Clarke was aged 35 in 1841 (i.e. born c.1806) and that he died in the first quarter of 1859.
69. E.M. Clarke, *List of prices of magnetical, philosophical, optical and chemical instruments and apparatus* . . ., London, 1838, 7.
70. P.J. McLaughlin, *Nicholas Callan: Priest-Scientist*, Dublin, 1965, gives the fullest account of his life.
71. P.J. McLaughlin, Nicholas Callan, *Dictionary of scientific biography*, **3**, New York, 1971, 18.
72. Described by Callan, On a new voltaic battery, cheap in its construction and use and more powerful than any yet made. And also a cheap substitute for nitric acid in the Grove platina battery, *London and Edinburgh Philosophical Magazine*, 3rd series, **31**, 1847, 81-85. Discussed and illustrated in *Exhibition illustrating some Irish contributions to the progress of the arts and science souvenir catalogue*, St Patrick's College, Maynooth, 1953, items 693-695; J.F. FitzMaurice Mills, *Encyclopaedia of antique scientific instruments*, London, 1983, 76; M.T. Casey, Nicholas Callan – priest, professor and scientist, *Proceedings of the Institute of Electrical Engineers*, **132** Part A, 1985, 496.
73. McLaughlin, *op. cit.*, 115-117 (ref. 70).
74. William Speer, *An enquiry into the causes of the errors and irregularities which take place in ascertaining the strength of spiritous liquors, by the Hydrometer, with a demonstration of the practicability of . . . rendering this instrument accurate*, London, 1802. Speer also took out a patent for his hydrometer in 1802, no. 2646.
75. G.H. Gabb and F.G.H. Tate, *Alcoholometry: an account of the British method of alcoholic strength determination with an historical introduction*, London, 1930, vii-xvii, 1-13.
76. *Ibid.*, xiv.
77. *Hibernian Journal*, 7 March 1810.
78. There is a ring dial in a private collection and a portable sundial was offered for sale at Sotheby's 1 November 1965 Lot 70.
79. William Speer, *Description of the use and application of a new patent saccharometer for ascertaining the gravity or wort, the produce of malt and the strengths of malt liquors etc.*, Dublin, 1806.
80. *Hibernian Journal*, 11 April 1806.
81. Inventory no. 1919-286; a copy of Speer's *Description . . ., op. cit.* (ref. 79) is also in the Science Museum, London, inventory no. 1876-480.
82. The following instruments with Spear's signature have been recorded: an ivory sector, British Museum, inventory no. 1906 11-5 1; a circumferentor, Dublin Civic Museum, inventory no. 223; a chrondrometer, National Museum of Ireland, inventory no. 4-1975; a sundial and an ivory sector, both National Museum of Ireland; a pair of opera glasses, Whipple Museum of the History of Science, Cambridge, inventory no. 2834; a universal inclining dial, Whipple inventory no. 819; a waywiser, Ulster Folk Museum; a small surveying sextant, a circular protractor and circumferentor are in a private collection; a microscope is in the Wellcome Collection at the Science Museum, London, inventory no. A76118; a theodolite and an ivory drawing scale are in the Depart-

ment of Engineering, Trinity College, Dublin; a pair of gunner's callipers are in the National Army Museum, London, and a clinometer, once belonging to Robert Halpin (1836-1894), commander of S.S. *Great Eastern*, is in the National Maritime Museum of Ireland. Besides these, a large variety of instruments has been offered at auction: a stick barometer, Phillips 4 March 1981 Lot 1; another, Christie's South Kensington 18 July 1985 Lot 1; a sympiesometer, Christie's 7 October 1981 Lot 1; a pocket sundial, Phillips 27 February 1980 Lot 76; another, Christie's South Kensington 13 March 1980 Lot 70; a drawing set, Phillips 26 October 1983 Lot 198; a clinometer, Phillips 2 February 1984 Lot 125; a theodolite, Sotheby's 25 October 1979 Lot 218; another, Sotheby's Belgravia 18 July 1980 Lot 251; a circumferentor, Sotheby's 31 March 1978 Lot 30; another, Christie's South Kensington 14 December 1978 Lot 242; a chart protractor, Phillips 21 May 1986 Lot 72; a telescope, Phillips 21 May 1986 Lot 105; a patent barometer, Sotheby's 19 July 1988 Lot 253; and a pair of gunner's callipers, Phillips 16 November 1988 Lot 106.

83. See Chapter Seven for more details.
84. J.H. Jellett, On a new optical saccharometer, *Proceedings of the Royal Irish Academy*, series 1, **8**, 1861-64, 279; *London, Edinburgh and Dublin Philosophical Magazine*, series 4, **32**, 1866, 357-360. Contemporary French literature called this instrument the "saccharimeter" (as it is universally known today), to distinguish it from the "saccharometer", which was the special form of hydrometer developed to measure the density of sugar solutions.
85. J.H. Jellett, A new analysing prism, *Proceedings of the Royal Irish Academy*, series 1, **7**, 1857-61, 350.
86. For more information on Bryson, see T.N. Clarke, A.D. Morrison-Low and A.D.C. Simpson, *Brass & glass: scientific instrument making workshops in Scotland as illustrated by instruments from the Frank collection at the Royal Museum of Scotland*, Edinburgh, 1989.
87. Jellett, *op. cit.*, 281 (ref. 84).
88. *The illustrated record and description catalogue of the Dublin International Exhibition of 1865*, London, 1866, 237.
89. *Ibid.*, 248-249.
90. *Dublin exhibition of arts, industries and manufactures, and loan museum of works of art*, 5th edn., Dublin, 1872, 88.
91. *Catalogue . . ., op. cit.*, 217 (ref. 19).
92. At the Museum of the History of Science, Oxford, inventory no. 36-13; and at the Department of Physics, Trinity College, Dublin. An incomplete cathetometer by Spencer & Son is in the Department of Physics, University College, Dublin.
93. A petrological microscope, a Norremberg polariscope and a Wollaston's goniometer are all in the collection of St Patrick's College, Maynooth; a polarising apparatus is at the Royal Dublin Society; a microscope is in the collection of the National Museum of Ireland, inventory no. 1-1975; another is in the Royal Museum of Scotland, inventory no. T1979.74 (described in R.H. Nuttall, *Microscopes from the Frank collection 1800-1860*, Jersey 1979, 50); another is illustrated in G. L'E. Turner, *Nineteenth century scientific instruments*, London, 1983, 168; and a French Oberhaeuser drum style microscope was offered for sale in Historical Technology, *Catalog* 131, Spring 1988, item 94.
94. G.L. Herries Davies, *Sheets of many colours: the mapping of Ireland's rocks 1750-1890*, Dublin, 1983, 198.
95. The Johnstone Stoney heliostat is discussed by A.A. Mills, Portable heliostats (solar illuminators), *Annals of Science*, **43**, 1986, 399-401. Examples are at the Whipple Museum of the History of Science, Cambridge, from the George Stokes Collection, inventory no. 1445; at the Museum of the History of Science, Oxford, used by the mineralogist M.H.N. Story-Maskelyne (1823-1911); at the Science Museum, London, used by the chemist William Abney (1844-1920), inventory no. 1921-106; and a fourth is in the Department of Physics, University College, Dublin.
96. *Cork Industrial Exhibition, 1803: report of executive committee, awards of jurors, and statement of accounts*, Cork, 1886, 379.
97. Mason, *op. cit.*, 148 (ref. 1).
98. One example is in the collection of the National Botanic Gardens, Glasnevin; another at the Royal College of Surgeons of Ireland.
99. Gillian B. Greenhill, The death of Chatterton, or photography and the law, *History of Photography*, **5**, 1981, 199-205.
100. Short histories of photography in Ireland have been written by Edward Chandler, *Photography in Dublin during the Victorian era*, Dublin, n.d. [1982] which had first appeared as a chapter in Tom Kennedy, *Victorian Dublin*, Dublin, n.d. [1980]; and Oscar S. Merne, *The story of the Photographic Society of Ireland 1854-1954*, Dublin, 1954. Much of the information given here comes from these works.
101. Wet plate camera at the Museum of the History of Science, Oxford.
102. Other suppliers of photographic apparatus in Dublin were: John Byrne (later John Byrne & Son), 1859-1900; F. Surridge, 1860-1876;

Richard Burton, 1862-1864; Thomas Mayne, 1880-1913; Gerald Mayne, 1914-1922.

103. *Proceedings for the election for Dublin . . . 1835*, Dublin, [1835], and *An alphabetical list of the constituency of the City of Dublin with the residence, qualification, and profession of each voter given at the elections of 1832, 1835 and 1837*, Dublin, [1837].

104. Frederick Walker, Householder, also voted against O'Connell, but it is not clear whether he was the instrument maker.

105. *Morning Post*, 11 June 1856. We are grateful to Project Simon for this information.

106. For instance, Joseph Cappo, who moved from Belfast to Glasgow and then returned. In general, the motives of migrants during the nineteenth century remain obscure: N.L. Tranter, *Population and society 1750-1940*, Harlow, 1985, 126-152.

107. M.V. Brewington, *Peabody Museum of navigational instruments*, Salem, Massachusetts, 1963, 135; Charles E. Smart, *The makers of surveying instruments in America since 1700*, New York, 1962, 107.

108. Project Simon has brought to our attention two examples from many others: George Hicks, mathematical instrument maker, Dublin, working 1820-1835; George Hicks, barometer maker, London, working 1810-1811; it is possible these were one and the same person.

109. Obituary, *Quarterly Journal of the Royal Meteorological Society*, **43**, 1917, 99.

110. Leonard G. Rule, *The flame of the lamp*, London, 1967, 9.

111. *Official descriptive and illustrated catalogue of the Great Exhibition 1851*, **1**, London, 1851, 446.

112. *Reports of the juries on the subjects in the 30 classes into which the exhibition was divided*, **2**, London, 1852, 544-545.

113. For a general account of the exhibition and an assessment of its impact see Alun C. Davies, Ireland's Crystal Palace, 1853, in J.M. Goldstrom and L.A. Clarkson (eds.), *Irish population, economy, and society: essays in honour of the late K.H. Connell*, Oxford, 1981, 249-270.

114. *Official Catalogue of the Great Industrial Exhibition (in connection with the Royal Dublin Society) 1853*, Dublin, n.d. [1853], 57-61.

115. Exposition des Produits de l'Industrie de tous les Nations, *Catalogue Officiel*, 2nd edn., Paris, [1855], 326-327.

116. International Exhibition of 1862, *The illustrated catalogue of the industrial department*, 4 vols, 2, London, 1862, 43.

117. F. Jenkin, Class XIII, Philosophical instruments and processes depending on their use, 37, in International Exhibition of 1862, *Reports by the juries on the subjects in the thirty-six classes into which the exhibition was divided*, London, 1863.

118. Paris Universal Exhibition, 1867, *Complete official catalogue*, 2nd edn., London and Paris, 1867, 196-197.

119. M. Lambert, Report on optical instruments, in the *Society of Arts: artisan reports on the Paris Universal exhibition of 1878*, London, 1879, 473-478. We are grateful to Dr Anita McConnell for drawing this article to our attention. Michael Lambert appears to have been in business on his own account c.1880, and an instrument signed by him – McCombie's Angle and Position Finder, 1898 – was offered for sale by Historical Technology, *Catalog*, 104, Spring 1972, item 181.

120. Lambert, *op. cit.*, 477-478 (ref. 119).

121. Howard Grubb, *Astronomical instruments, observatories etc. manufactured by Sir Howard Grubb FRS, FRAS etc.*, Dublin, 1908, 8. We are grateful to Dr Jon Darius for this reference.

122. Dublin Exhibition of Arts, Industries and Manufacturers, *Official catalogue*, 5th edn., Dublin, 1872, 38-40.

123. International Inventions Exhibition, *Official catalogue*, London, 1885, 267.

124. Thomson *v.* Moore, in John Cutler (ed.), *Reports of patent, design, and trade mark cases*, **6**, 1890, 426-464. Our thanks to Dr Anita McConnell for this reference.

125. International Inventions Exhibition, *op. cit.*, 48 (ref. 123).

126. Patent no. 9969, 22 August 1885.

127. International Inventions Exhibition, *op. cit.*, 264 (ref. 123).

128. J.Y. B[uchanan], Walter Noel Hartley, 1846-1913, *Journal of the Chemical Society*, **105**, 1914, 1207-1216.

129. Inventory no. 1914-323.

130. International Inventions Exhibition, *op. cit.*, 265 (ref. 123).

131. John Joly, On an apparatus for obtaining telegraphically the readings of meteorological instruments placed at a distance from the observer, *Scientific Proceedings of the Royal Dublin Society*, **4**, 1884, 158-165.

132. John Joly, On a method of determining the specific gravity of small quantities of dense or porous bodies, *ibid.*, **5**, 1886, 41-47.

133. John Joly, On photometers made of solid paraffin, or other translucent substance, *ibid.*, **4**, 1885, 345-346.

134. John Joly, On the determination of the melting points of minerals. Part 1. The uses of the meldometer, *Proceedings of the Royal Irish Academy*, 3rd series, **2**, 1891--93, 38-64.

135. George Francis Fitzgerald, On a non-sparking dynamo, *Scientific Proceedings of the Royal Dublin Society*, **4**, 1885, 343-344.
136. George Johnstone Stoney, On the possibility of prolonging the tones of a pianoforte so to produce an instrument with the quality of tone of the piano, and the continuity of tone of the organ, *ibid.*, **4**, 1883, 147-148.
137. Casserley, *op. cit.*, 220-222 (ref. 34).
138. C.R.C. Tichborne, On an argentiferous galenetic-blende at Avoca, *Scientific Proceedings of the Royal Dublin Society*, **4**, 1885, 300-305, quoting from p. 300.
139. B.B. Kelham, The Royal College of Science for Ireland (1867-1926), *Studies*, **56**, 1967, 297-309.
140. *Irish Times*, 4 April 1866.
141. Royal College of Science for Ireland, *Directory . . . 1881-82*, Dublin, 1881, 4.
142. W.F. Barrett, On a new form of calorimeter, *Scientific Proceedings of the Royal Dublin Society*, **5**, 1886, 13-16.
143. W.F. Barrett, On the physical properties of manganese steel, *ibid.*, **5**, 1887, 360-371.
144. James Parkes & Son, *Wholesale catalogue of optical, mathematical and philosophical instruments*, Birmingham, 1867, 1. Copy in Birmingham Central Library.
145. James Parkes & Son, *Wholesale catalogue of microscopic apparatus*, 2nd edn., Birmingham, 1886, 1. Copy in Birmingham Central Library.
146. R. & J. Beck, *The "London" microscope*, London, 1900, 2. Copy in Museum Boerhaave, Leyden.
147. In a private collection.
148. University College Dublin, Archives, RCSI/263.
149. Yeates & Son, *op. cit.*, 36 (ref. 31).
150. University College Dublin, Archives, RCSI/262.
151. M.J.G. Cattermole and A.F. Wolfe, *Horace Darwin's shop: a history of the Cambridge Scientific Instrument Company 1878 to 1968*, Bristol, 1987, 54-56.
152. Joseph M. Maiben, *Illustrated and descriptive catalogue comprising chemical, biological and bacteriological apparatus, lecture and general elementary physical apparatus, balances and weights...*, Dublin, 1913. Copies in the Royal Dublin Society and the Royal Museum of Scotland.
153. *Ibid.*, passim.
154. *Ibid.*, 24-37.
155. *Ibid.*, 415.
156. *Ibid.*, 97.
157. *Ibid.*, passim.
158. *Ibid.*, 211.
159. *Ibid.*, 330.
160. For example, T.D. Spearman, Mathematics and theoretical physics, in T.O. Raifeartaigh, (ed.), *The Royal Irish Academy: a bicentenniel history 1785-1985*, Dublin, 1985, 200-239; G.L. Herries Davies, Astronomy, geology, meterology, in *ibid.*, 247-273.
161. John F. Fleetwood, *The history of medicine in Ireland*, 2nd edn., Dublin, 1983, 132.
162. Valentine Ball, A geologist's contribution to the history of ancient India, being the presidential address to the Royal Geological Society of Ireland, *Scientific Proceedings of the Royal Dublin Society*, **4**, 1883, 69-116.
163. Letter of G.H. Kinahan to the Committee on Science of the Royal Irish Academy, 31 May 1890. Royal Irish Academy, MS Minutes of the Committee on Science, vol. 3, 364-365.
164. Robert Lloyd Praeger, *The way that I went*, Dublin, 1937; and *Some Irish naturalists*, Dundalk, 1949.
165. G.L. Herries Davies, The earth sciences in Irish serial publications 1787-1977, *Journal of the Earth Sciences Royal Dublin Society*, **1**, 1978, 1-23.
166. G.L. Herries Davies, Irish thought in science, in Richard Kearney (ed.), *The Irish mind: exploring intellectual traditions*, Dublin, 1985, 292-310.
167. *Dublin Express*, 28 November 1866.
168. Mary Laffan, *Hogan, M.P.*, Dublin, 1881, 195.
169. Maxwell Simpson, Presidential address to Section B, *British Association Reports*, 1878, 501-504, quoting from p. 501.

CHAPTER FIVE

IRISH PROVINCIAL INSTRUMENT MAKING

J.E. Burnett and A.D. Morrison-Low

Instrument making in Cork and district

Cork has been the most important trading centre for the south of Ireland for many centuries. Until the rapid growth of Belfast in the nineteenth century, Cork was the second largest centre of population in Ireland; it is now the second city of Eire. Cork harbour provides a large secure anchorage – thus the traditional boast that it could accommodate the whole of the British fleet – but it only achieved its greatest status when steam ships began to ply across the Atlantic. Steamers were able to enter the harbour in all winds, and when the *Sirius* made the first steam crossing to America in 1838, she left from Cork.

The first known Cork instrument makers are Elias and Daniel Voster, father and son. From his name it would appear that Elias Voster was an immigrant. This impression is strengthened by the knowledge that he married a woman with a Dutch name, Joanna Schuurman, in Ireland in 1702[1]. He was a teacher of mathematics. The preface to his *Arithmetic*, last reprinted in 1843, says that the book "contains such Questions in most of the Rules of vulgar ARITHMETIC, as I have taught my *Scholars* for many years past . . ."[2]. The date of Daniel Voster's birth is not known, but he married Ann Morris in 1737[3]. Judith Voster, perhaps his sister, had married

5.1. Detail of wooden Gunter's rule signed "Daniel Voster Corke Fecit 1742". Trustees of the Science Museum, London.

Isaac Bussy in 1735[4]. Daniel died in May 1760. The *Cork Evening Post* reported, "last Saturday died at his house in Brown Street after a few days illness Mr Daniel Voster . . . [he had an] extensive knowledge of the several branches of the Mathematicks . . ."[5]. The *Dublin Gazette* added "died at Corke, Mr Daniel Voster an eminent teacher of Mathematics"[6]. The following month an advertisement appeared[7]:

> Notice is hereby given
> That the WIDOW VOSTER in Brown-Street, having determined to quit the business, followed by her late Husband, will sell at very reasonable rates, several curious Instruments for the Use of Mariners, Surveyors, &c. In the mean time constant attendance, will be given to mend and repair all kinds of Mathematical Instruments; the person employed by the late Mr Voster, will be continued for that purpose.

Two separate schools were to be established to continue Voster's teaching of mathematical subjects. The level of mathematical activity in Cork is suggested by advertisements placed by no fewer than four other teachers in the same copy of the *Cork Evening Post*. Three months later Mrs Voster placed a notice in the *Cork Journal*: "Mrs Ann Voster in Brown street advertises the sale of mathematical instruments etc"[8].

The other family who operated in Cork during the second half of the eighteenth century were the Sweenys. Four of the family are known: Nathaniel (d. 1763), Edward (fl. 1763), Widow Sweeny (fl. 1795), and Edward Nathaniel (fl. 1798). Edward took over his father's business on the latter's death in 1763[9]. In addition to making mathematical instruments, Edward Sweeny advertised that he carried out copper-plate printing. An octant, signed "Sweeny Cork Fecit" is in the collection of the National Maritime Museum, Greenwich (*Fig.* 5.2)[10]. The last of the family issued an advertisement in March 1798. It indicates that a wide range of instruments was for sale, both for the professional and the dilettante customer[11]:

> Edw. Nathaniel Sweeny, Mathematical and Optical Instrument Maker, No. 27 Patrick Street, CORK
>
> Makes and Sells Barometers and Thermometers, engaged to equal any made in this kingdom. Theodolites, Circumferentors, Electrifying Machines, Concave and Convex Glasses, mounted in gold, silver Tortoiseshell and Horn; Spectacles of various kinds; Hydrometers, Sextants, Quadrants, Compasses, Cases of drawing Instruments; Sun Dials, Telescopes, Opera Glasses, Linnen [sic] Provers, &c.
> Every Article in the above Business repaired with accuracy and dispatch, and on the most reasonable Terms.
> An extensive assortment of Sea Charts and Pilots.

Cork was then at its commercial zenith. The population is estimated to have grown from 56,000 in 1730 to 80,000 in 1775[12]. In the middle of the seventeenth century it had been the fifth city in Ireland, but now it was the second[13]. A formidable export trade in butter had been built up – it lasted until the second half of the nineteenth century. Cork's expansion, indeed, predates the Industrial Revolution. It drew butter from all over the southern half of Munster, and there was a supporting trade in barrelled beef[14]. Cork was the last port of call for shipping from Europe to North America and the Caribbean, and it enjoyed a handsome trade in provisioning those vessels. Between 1776 and 1800 provisioning was at its peak, partly because extensive supplies were being bought by the Royal Navy[15]. Cork was both wealthy and cosmopolitan. A single page of the *Cork Evening Post* advertised, "clarets of the principal growths of Medoc Vintages 70 and 71, Frontiniac, Preniac, genuine old carcalla . . . Jamaica and Barbadoes rum . . . Malaga in quarter casks . . . geneva in pipes or hogsheads . . . Glo'ster cheese . . . Wiltshire beer, Hereford cider, Durham flower mustard . . ."[16]. The influx of foreign spirits must have

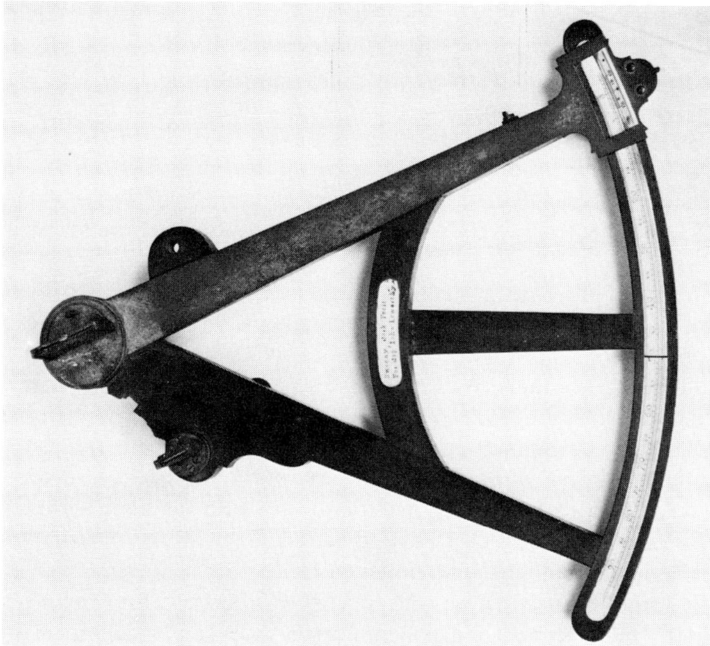

5.2. 15-inch octant signed "Sweeny, Cork", c.1780. National Maritime Museum, Greenwich.

quickened the need for Sweeny's hydrometers. Yet Edward Sweeny's advertisement, despite the overall prosperity of Cork, was badly timed. A few weeks later the rebellion of 1798 started.

Business was at this period sufficiently large to attract John Booker, who had a large trade as a looking-glass maker in Dublin, though it is known that he sold some instruments also (see Chapter Three). For example, he advertised in August 1773 that he would be in Cork on the Monday before the next Assizes with a supply of his goods[17]. Balthazar Knie, the well-known Edinburgh barometer maker, declared that he worked in Cork before he came to Scotland. German in origin, Knie was active in Edinburgh between 1776 and 1815[18]. In 1793 he said he had been making barometers "in various parts of Europe" for over forty years[19]. Westropp records a man named Knie working in Cork in 1773[20] and, although we have been unable to trace Westropp's source, there is a corroborative letter in the *Hibernian Chronicle*, a Cork newspaper, for 18 January 1773:

> Mr Flyn, I have for many years past made observations on the rise and fall of the quicksilver in the Barometer, but did never in the most violent storm, or at any other time see it so low as it was on last Sunday. About 10 o'clock, at night it was lower by the 10th of an inch than 28 inches and from its flat surface on the top seem'd to be still falling. As there was nothing remarkable in the weather at that time it surprized me having never heard of its being seen so low. Some of your learned Correspondents will perhaps account for this extreme lightness of the Atmosphere and what it portends.
> I am, Sir, your frequent Reader
> BALTHASAR WEATHERWISE

Balthasar Weatherwise! – the evidence is sufficient for us to be certain that this is a pseudonym of Knie's.

In tracing the history of instrument making in Cork, we are hampered by the lack of a street directory in continuous publication before 1875. The first directory was issued in 1787, and it reveals two makers at work, Alexander and Francis Fitton, both in Paul Street[21]. The Sweenys do not appear. Francis Fitton's description in the directory as a "Dealer in Glass and Earthenware, Ship-Chandler, and Mathematical Instrument-maker"[22] suggests that he may only have been a seller of instruments, but the existence of an octant in the Peabody Museum, Salem, Massachusetts, marked "Made by Francis Fitton Cork For Captn Abel Orpin 1784" appears to give the lie to this[23]. He was almost certainly the Francis Fitton who died in 1801[24], and may be related to Nicholas Fitton, brass and copper founder of Tuckey's Quay, Cork, who advertised in 1753 and died in 1781[25]. Francis Fitton is known to have taken an apprentice in 1788[26].

As with so many of the Cork instrument makers, none of Alexander Fitton's products is known to survive, so it is impossible to offer any comment on his work. Nor can we be sure of the scale of his operations, since he also sold musical instruments[27], although this was not so unusual as it would seem, since James Watt made and sold both musical and mathematical instruments in Glasgow in the 1760s[28]. Michael Fitzgerald, who in 1820 was known as a land surveyor and mathematical instrument maker, is another man who carried out varied activities, though his obituary mentioned only that he was a land surveyor[29], and in the directory for 1812, and in one of the two for 1820, he is indeed listed only as a land surveyor[30]. He is known to have been active as a land surveyor between 1815 and 1824[31].

The number of instrument makers known to be trading in Cork remained steady at two or three from 1787 until c.1875. Those who remained in business for the longest periods were the Hunt family (fl. 1805-1893), Thomas Bennett (fl. 1809-1867) and Mrs Margaret MacHugh (fl. 1863-1883). Obviously, it is likely that other shopkeepers sold small numbers of instruments as a part of their particular trades. The partnership of Edwards & Savage, booksellers, did so: they also sold "Gunter's Scales and Dividers; and Mathematical Instruments in Cases of various sizes"[32].

After the Great Exhibition of 1851, it was widely felt that Ireland had been under-represented in the Crystal Palace, and to compensate a National Exhibition was held at Cork the following summer. The Mayor of Cork wrote a synopsis of the exhibition, and said, "Thomas Bennett and Henry Hunt, both of Cork, have exhibited conclusive proof of their ability, as makers of Nautical, Mathematical and Philosophical Instruments"[33]. Lest the reader should have any doubt about the skill of Bennett and Hunt, the Mayor added, "it is a fact, curious but true, that the best compasses, almost in the world, are made in three places, Cork, Liverpool and Trieste"[34]. Thomas Bennett appears to have been the more enterprising: for a time around 1863 he had, in addition to his shop in Patrick Street, a branch at Queenstown[35]. The trade offered by the ships using the deep water berths there does not seem to have been as great as he had hoped (Queenstown had become a regular port of call for trans-Atlantic shipping in 1859), and no further references to the selling of scientific instruments there are known.

There is clear evidence that Bennett had a high degree of manipulative skill. An advertisement of 1852 reads, "T. Bennett, optician, manufacturer of mathematical, optical and philosophical instruments, 124 Patrick Street, Cork. All orders for NEW WORK promptly attended to, and REPAIRS executed with carea nd [sic] punctually"[36]. Bennett's advertisement in 1863 described him as an "Adjuster of Compasses in Iron Ships, and manufacturer of Mathematical, Optical and Nautical Instruments"[37]. Bennett was said to have made improvements to instruments "which practical men think of the higher value", and to have invented some new manufacturing tools[38]. It is conceivable that he was the Thomas Bennett who was apprenticed to William Stokesbury of Cork, brassfounder, 1791, aged about 15[39], but it is also difficult to believe that Bennett had not been trained in London.

It is possible to judge the quality of Bennett's work from a number of surviving instruments at University College, Cork. A disc electrical machine with a 24-inch disc, signed on an ivory label "THOS. BENNETT/Cork" is clearly the work of a skilled instrument maker. The collection

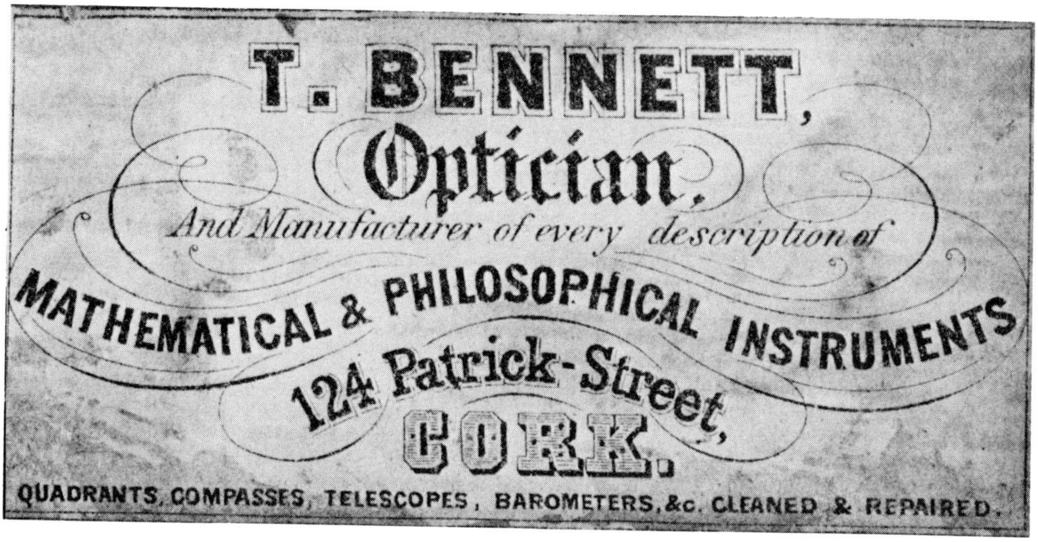

5.3. Trade card by Thomas Bennett, Cork, mid nineteenth Century. University College, Cork.

there also has a number of other pieces of electrostatic apparatus which are of the same quality and style. They, too, may well be by Bennett. A small improved sympiesometer at University College, Cork, bears Bennett's name, and suggests that he may have had some skill in glass-blowing. Unfortunately an ebony octant dating from 1832-1859 has in its box merely Bennett's label (*Fig.* 5.3), presumably as repairer, the instrument itself having been signed by Buckley of Dublin[40]. As his advertisements suggest, much of Bennett's trade was in instruments for the professional market. In 1852 he exhibited dumpy and irrigational levels, ships' binnacles and a variety of compasses, marine and domestic barometers, thermometers, vacuum gauges, and steam gauges, tubular timber callipers, as well as "an electro-magnetic machine" and other apparatus[41]. Bennett also took part in the activities of local scientific societies. At a conversazione of the Cork Cuverian Society in the Athenaeum in 1855, he performed optical illusions, and displayed an aneroid barometer, a Bourdon steam pressure gauge and an apparatus for demonstrating the precession of the equinoxes[42].

Other instruments known to have been sold by Bennett are a compass-dial[43], a stick barometer[44], an octant[45], and a quintant[46]. The last two have ebony frames. Also a sextant, a Sikes hydrometer, a clinometer, an artificial horizon, a Henley electrometer, a Coulomb electrometer and a wet and dry bulb thermometer have been recorded[47]. Bennett's label is also found on the back of a copy of J.S. Hobbs, *A Chart of the Cattegut, the Baltic or East Sea* (1858)[48].

It seems likely that there was slow expansion of instrument making in Cork between 1800 and 1850. Not only was science itself expanding, but also it was bringing new institutions with it. There may even have been an unsatisfied demand for apparatus at the turn of the century. Thomas Dix Hincks, who gave a course of public lectures in 1803, which included many demonstrations, almost as soon as similar lectures had been instituted by James Lynch in Dublin, complained that "the lecturer met with several disappointments with the respect to the apparatus necessary for illuminating many subjects"[49]. This would have been unlikely if he had had access to a skilled instrument maker.

The Royal Cork Institution was founded in 1807, with professors of chemistry, natural philosophy, and agriculture and a lecturer in natural history. Modelled partly on the Royal Institution in London, and partly on the Royal Dublin Society, the Royal Cork Institution flourished until government funding was withdrawn in 1831, and then died a lingering death over the next half century[50]. In its early years it bought a certain amount of scientific apparatus, including a gazometer made by "our townsman, Mr Fitzgerald"[51]. Though one suspects that most of the demand for instruments in Cork in the first half of the nineteenth century was for navigational items, it is interesting to note that the largest number of instrument makers to have traded simultaneously in Cork – four – occurs during the more active period in the life of the Royal Cork Institution. In 1812, Thomas Bennett, Francis Geary, Thomas Hunt and Michael Walsh were in trade[52]; in 1817, Bennett, Geary, and James and John Hunt, trading from different addresses[53]. In 1819 the Cork Scientific and Literary Society came into being. It was followed c.1836 by a parallel body, the Cork Cuverian Society. Most significantly, the incorporation of Queen's College, Cork in 1845 (changed to University College, Cork, in 1908), brought to the city an institution offering higher science education. It admitted its first students in 1849. The medical school had a particular place in the College. Between about 1860 and about 1920 at least half of the students at University College, Cork – and sometimes more than three-quarters of them – were reading medicine[54]. Natural philosophy was a compulsory subject, and so there was heavy demand for demonstration apparatus.

Thomas Bennett last appears in a directory for 1867[55]. By 1870 his premises were in the hands of Messrs Reynolds & Wiggins[56], and they appear to have taken over Bennett's business, since in 1872 directory lists them as "Reynolds & Wiggins (late Bennett)"[57]. This, however, was the last appearance of this short-lived firm. The shop at 124 Patrick Street was taken over by Purcell, the publisher of the *Commercial Cork Almanac*.

Thomas Hunt, the elder, is said to have opened his shop at 118 Patrick Street in 1792[58]. After his death c.1815, the business was successively continued by his two sons, Thomas (died c.1835) and Henry. The latter's name was used until c.1885, but it is not clear whether the firm was taken over by a member of the third generation at that point, or (more likely) earlier. As with Bennett, the market they served was largely professional. In the 1852 exhibition they "exhibited a full mounted five-inch theodolite, with all the adjustments; a fourteen-inch dumpy level; a circular protractor with double verniers, silver arc; a three-and-a-half-foot, fullsized achromatic telescope, with vertical and horizontal rack-work motion; a transparent dipping compass; also, a patent metallic dipping compass; besides binnacles of various kinds"[59]. Forty years later navigational and surveying instruments were the Hunts' main products, though they also made spectacles[60]. To the end they had workshops of their own. Three years before the business closed, a description of their shop stated that "spacious workshops are situated in the rear of the premises, and on the floor above a number of highly skilled workmen are employed in the manufacture and repair of the numerous nautical and other instruments"[61].

From the little evidence available it appears that the firms of Bennett and Hunt were roughly the same size; Griffith's valuation (1852) placed a net annual value of £45 on Bennett's premises and £40 on Hunt's[62]. Few instruments made by Hunt are known. A three-draw telescope signed "H. Hunt/Cork" in a private collection, a marine barometer signed "T. Hunt Cork" and a wheel barometer signed "Thos. Hunt" have been recorded[63].

After 1875 there are difficulties in distinguishing makers (or indeed sellers) of scientific instruments in Cork, from the ophthalmic opticians who also often describe themselves (or are described by directories) purely as opticians. In some cases the distinction is clear. For example, P.J. Fielding "chemist-optician" advertised that he was a pharmacist who in addition sold spectacles[64]: clearly he did not trade in instruments. John Gilbert, at 120 Patrick Street for many years from c.1875, generally appears in directories under the heading "Opticians", but with the description "Fine Art Warehouse". On one occasion a more elaborate description was provided: "Fine Art Repository, Optician to the Faculty, and to the Ophthalmic Hospital, Frame Maker, Illuminator, etc."[65]. He, and others like him, are not included among the Cork instrument makers listed in Chapter Eight. Some others, such as William Barriscale, and Hilser

Brothers, who were primarily watchmakers or jewellers, may have sold telescopes, binoculars or barometers, but they have also been excluded from the list in the absence of more evidence[66].

Despite these problems of definition, it is certain that instrument making (strictly speaking) stopped in Cork well before 1921. The most likely point at which it ceased is the demise of the venerable firm of Henry Hunt – by then known as Hunt & Son – in 1895[67]. It is highly likely that their business had been killed by the larger Dublin manufacturers – just as theirs, in turn, was eclipsed by British, European and American competition. A particularly graphic account of Irish attitudes was given by John Francis Maguire, and it is so important that it is quoted here *in extenso*[68]:

> Although there is no small amount of engineering done in Ireland, principally caused by the construction of railways, still the absurd prejudice is so great – of Irishman against Irishman – that it is with difficulty, in many instances, that an Irish maker of the necessary scientific apparatus can secure sale for the product of his mechanical skill and practical science. Even the newest boy, who has just got his first job on a railway, disdains the work of a man who has ten times his knowledge, and will not be satisfied unless he procures his dumpy and his levelling staff from Mr. So and So of London; and even professors in Irish colleges and public schools, who should be the very first to promote the manufacture of scientific apparatus, in their own country, make it imperative on their pupils to order their instruments in London. I have heard a reason given by a customer of a local manufacturer for ordering, through him, a London-made article, and refusing to purchase one which the Cork maker offered at considerably less than the London price, and which he was willing to submit to any possible test – "Oh, I would not buy one of yours at all; for if I were done with it tomorrow, I could get nothing for it – the *name* would destroy its value". Allowing this true anecdote to tell its own tale, I may merely remark, that the man who can produce such an instrument as the theodolite exhibited by Mr Hunt, could make anything in the mathematical line; and that all that such a man wants, to extend his trade, and to add to the employment of the country, is fair play, and no favour. Test his work after what fashion you please, and judge him by the result; but do not unjustly condemn him to be the victim of false and degrading prejudice; do not doom your own countryman to a hard struggle with the tax-collector, while you make the fortunes of strangers . . .

Maguire was not merely encouraging the Irish people to buy Irish scientific instruments, he was trying to stimulate local industry by promoting the purchase of all forms of Irish goods. This, indeed, had been the aim of the 1852 exhibition[69]. Through Maguire we have a direct link between the manufacture of scientific instruments and nationalist politics in Ireland. He founded the *Cork Examiner* in 1841, to support Daniel O'Connell. In parliament from 1852 until his death 20 years later, he fought for tenants' rights and pressed for the disestablishment of the Church of Ireland[70].

Some instruments obviously continued to be sold in Cork after the demise of Hunt & Son. For example, W. & T. Evans, booksellers at 11 Patrick Street, in 1914 advertised drawing instruments[71].

As a footnote to the history of instrument making in Cork we must consider the Kinsale magnet (*Fig.* 5.4). Kinsale is a small coastal town some 20 miles south of Cork. In the seventeenth century it was of far more importance than it is today, and for 20 years after the Restoration it was the leading port on the south coast of Ireland. It is quite possible that an instrument maker practised there in the period between 1660 and 1680, although there is no evidence that one did so: the market, however, was there. By the nineteenth century Kinsale was slumbering, its forts overgrown. In the Whipple Museum, Cambridge, is a large permanent magnet in a brass mounting, engraved "W. Lewis FECIT Kinsale 1843"[72]. It is well made and the engraving is neat and regular. William Lewis seems to have been an amateur scientist. He appears in a variety of direc-

5.4. Magnet, by William Lewis of Kinsale, 1843. Whipple Museum of the History of Science, Cambridge.

tories between 1844 and 1870, under the heading "Gentry and Clergy": most likely he was a minor landowner[73]. In Griffith's valuation his address is given as 34 Compass Hill[74].

Instrument making in Belfast and Ulster

The growth of Belfast took place much later than that of the other major Irish cities; indeed, it was the last city in these islands to reap the benefits (and suffer the disadvantages) of the Industrial Revolution, and by the end of the nineteenth century its prosperity lay with its textile, shipbuilding and engineering trades[75]. Initially, most of Ireland's export trade had passed through Dublin, where merchants were able to find credit facilities more easily; but the agricultural products which at first made up a large portion of Belfast's exports gradually gave way to linen, and by the 1770s the city shipped over one fifth of Ireland's linen exports[76].

During the eighteenth century, Belfast's population doubled to 20,000[77], "grew from 37,000 in 1821, to 75,000 in 1841 and to 387,000 in 1911, a remarkable expansion by the standards, not merely of Ireland, but of the United Kingdom as well. In 1821 barely 2 per cent of Ulster's population had lived in Belfast; in 1911 the proportion was almost one quarter"[78]. Over the nineteenth century, there were three main demographic trends in Ulster, which contrasted with the straightforward pattern of growth in Belfast. "First, the population of the province, after rising from 2 million to 2.4 million between 1821 and 1841, fell to 1.6 million in 1911. Second, Ulster's share of the population of Ireland increased from 29 to 36 per cent. Third, the proportion of the population in the province living in towns grew from less than 10 per cent to almost 40 per cent"[79]. This is in strong contrast with Dublin, the population and industries of which were in decline by the end of the century[80]. Yet despite the growth of the port, and the gradual change of the nature of Belfast from a market town to a commercial centre to a manufacturing city, it never became a provincial centre for instrument making, as did, for instance, Glasgow, with which Belfast had much in common[81].

Most of the trade in scientific instruments in the Belfast area appears to have been a retail trade, with only the isolated maker apparently managing to make a living, and then only by making other articles as well. On the whole, items were bought in from London or Dublin, cities which were large enough to support their own instrument making communities – which Belfast was not, until the late nineteenth century, by which time the nature of the trade had changed from a craft-based business to factory production[82]. The bought-in instruments were then sold by shopkeepers who often did not even pretend to make all the items that they sold. For example, a hardware advertisement, placed in the *Belfast News-Letter* in 1764, gives some idea of the nature of the trade[83]:

> John Bashford, who served his Apprenticeship to Mr William Hillditch, has just opened the Shop in High Street Belfast, lately occupied by Henry and Robert Joy; having imported from London, Birmingham and Sheffield, and other Places, a neat Assortment of HARD-WARE, carefully collected by himself on the Spot; consisting of the following articles: viz. [here follows a range of ironmongery of mindboggling variety] . . . Temple Spectacles . . . Guns and Pistols . . . Magnets; Carpenters, Shoemakers, Coopers and Other Mechanic Tools; inside Mainsprings, and Chains, Keys and enamelled Dial-Plates for Watches; Furniture for Coffins; Beams and Scales . . . Brass and Iron Compasses, Scales for Mariners; best Box sliding Rules; best common Ditto . . .

Not every merchant travelled to England to collect his stock. Thomas Ward, "Bookseller, Bookbinder, and Stationer in Lisburn" advertised in 1783 that "he having lately returned from Dublin, has laid in a large and complete assortment of the following articles . . . mathematical instruments, joiner's rules, balances for weighing gold, good money scales and weights, . . ."[84]. Others waited until ships arrived in Belfast, laden with fancy goods. Amongst the lists of hardware was: "Thomas Lyle, at the Saw, in High Street, Belfast, Hath lately imported from England and Holland &c . . . Cases of Instruments for Drawing, Gunters scales and dividers; English, Irish and Dutch Weighing Beams, Money Scales"[85].

From this evidence it would appear that there was a very small market for instruments in Ulster in the eighteenth and early nineteenth centuries. Occasionally items were offered for sale, second hand, which would appeal to the dilettante market, such as the "most compleat three pillar'd microscope" offered for sale in 1760, together with a London-made lathe, by William Mathers, a watchmaker from Ballycastle[86]; or the "curious reflecting Telescope" at a booksale in 1774[87]. There seems little evidence at this period for indigenous production of either surveying or nautical instruments: broadly speaking, it would appear that Dublin was able to supply the whole of Ireland with the former, while England could manage the latter. Captain Francis Dubourdieu (?1784-1861), of the Royal Hanoverian Engineers, from a Huguenot family who had settled in Lisburn, owned a plain theodolite which had been made by Lynch of Dublin around 1800[88]. There was a very limited market for teaching and demonstration apparatus, despite the founding of the Belfast Natural History Society in 1821 (renamed the Belfast Natural History and Philosophy Society in 1842)[89], local taste was more for natural history, than for natural philosophy as the *Proceedings* show[90]. The Belfast Reading Society, which later became the Belfast Library and Society for Promoting Knowledge (commonly known as the Linen Hall Library) was established in 1788 with "the object of . . . the collection of an extensive Library, Philosophical Apparatus, and such productions of Nature and Art as tend to improve the mind, and excite a spirit of general enquiry"[91]. Their first instrument, an air pump, appears to have been bought in March 1793, but probably came from a London source[92]. By 1805 the Society possessed a pair of globes, an air pump, a rain gauge, a Wedgwood's pyrometer, two barometers and two thermometers[93]. Itinerant lectures, certainly in the eighteenth century, were

5.5. Wheel barometer with paper scales, signed "John Noseda fecit". Courtesy of Sotheby's, Sussex.

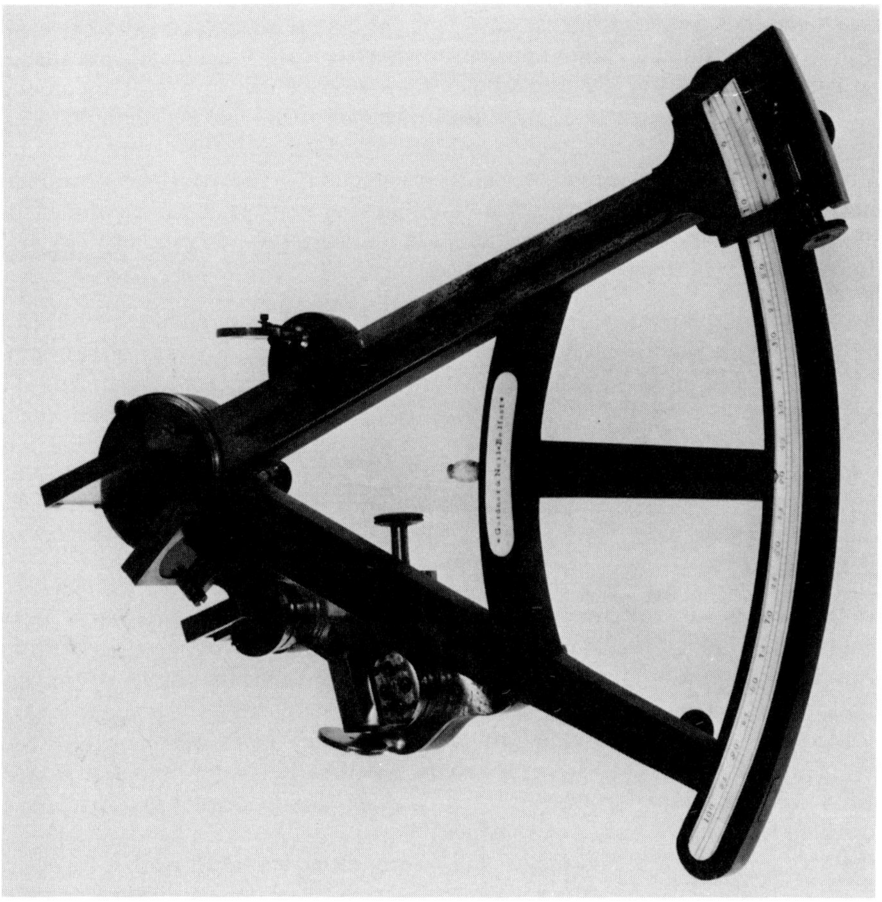

*5.6. 10-inch octant, which once belonged to Lord Kelvin's father, signed "Gardner & Neil * Belfast". Department of National Philosophy, University of Glasgow.*

few and brought their own equipment with them. When accidents happened, as befell a Mr Walker from London in 1767, help had to be sought from outside Ulster: "A. Walker is sorry to inform his present subscribers, that by having broke the Cylinder of his Electrical Machine in the Carriage, he is obliged to misplace that lecture and wait till one he has sent for from Dublin arrives, which he hopes will be in a few Days"[94]. The market, such as it was, was restricted by the small population, which was still essentially a rural one, and whose needs for scientific instruments were fairly basic.

The first evidence of a Belfast-based instrument maker, as opposed to a retailer, comes from an advertisement placed in the *Belfast News-Letter* in 1774. This is addressed "to linen drapers", and in it John Noseda (probably an Italian) weatherglass maker, advertised "that he has invented a kind of Balls for trying the Strength of Sour used in Bleaching, on the same Principle of those for trying spirituous Liquors, which have been found to be infallible, and of Course may be of singular Advantage in that Art, Linens having frequently been materially damaged by the Strength of the Sour not being properly ascertained". These sets of philosophical bubbles, deliberately aimed at a local market, came with printed instruction for use in two forms: "Boxes containing three Balls, necessary to try the different kinds of Sour, 6s 6d – Boxes, containing two Sets of six Balls at 11s 4d h[alf penny]"[95]. There was no repetition or follow-up to this advertisement, and Noseda died some five years later, so it is impossible to gauge how successful his "Balls" were[96].

A later maker of "hydrostatical glass beads" was George Myers, who advertised them in 1810 "for proving Spirits, Malt Liquor, Bleaches; Lyes and Dyes, and all kinds of Waters . . . N.B. The above are carefully adjusted by himself, and will be sold Wholesale and Retail"[97]. He went on to say that he was selling off his entire stock of glass, china and earthenware, presumably to concentrate more upon the specialist glassware side of his business: later directories show that he described himself as a "Barometer and looking-glass manufacturer"[98]. Others involved in this sort of business at the start of the nineteenth century were people with names (like John Noseda) which suggest that they were of Italian extraction: Anthony Capo or Cappo and G. Gioccomelli are two examples. The former settled in Belfast, and is known from entries in street directories (although his near-relation – a brother, or perhaps a son? – Joseph Cappo, seems to have settled in Glasgow for a time[99]), whereas the latter is recorded only through his instruments, and may well have moved around[100].

Another glassworker, albeit part-time, was Richard Cuttler, watchmaker, engraver and barometer maker, who worked "opposite the Castle-gate, Carrick-fergus" in 1782[101]. He may have been the Richard Cutler (sic) recorded as a watch and clockmaker, engraver and "Barramatition" at Youghal in 1787[102]. Cuttler claimed to have learned barometer making "which he purchased at a dear rate up the Mediterranean", so he may well have practised his craft in a number of places[103].

The watch and clockmaking and selling trade appears to have overlapped considerably with the instrument making and selling community throughout the eighteenth and nineteenth centuries. For instance, of three Armagh brothers involved in the clockmaking trade, John, George and James Waugh, the last named was also involved in producing astronomical instruments for the newly-founded Armagh Observatory in 1793[104]. Most of the instruments were produced by London makers, such as Dollond and Troughton, but "the transit instrument was made by Mr James Waugh, a very ingenious mechanic, and a native of Armagh", as was "an astronomical quadrant of 35 inches radius, constructed by the Rev. Robert Hogg, with the assistance of James Waugh, above-mentioned"[105]. Hogg later became assistant to the Astronomer at Armagh. One is strongly reminded of John King, assistant to Professor Patrick Copland of Marischal College, Aberdeen: both worked together on the construction of an astronomical clock in the 1780s[106].

Another late eighteenth century watch and clockmaker who was mechanically gifted and seems to have undertaken some instrument making, was Job Rider, who came to Belfast in 1791 and announced his arrival in the local press: "Job Rider, from London and Dublin, and last from Hillsborough, has commenced business at the sign of 'The Reflecting Telescope' at Shambles Street, where he makes clocks and watches of all kinds . . ."[107]. The following year he made a barometric clock, and appears to have made optical instruments too. He also repaired the Belfast Reading Society's air pump shortly after its purchase in March 1793[108]. In 1807 he ended a two-year partnership with Henry Gardner, and joined William Boyd in setting up the Belfast Foundry. This partnership was dissolved in 1831 and Rider started up a similar business of his own, but died on 4 April 1833 at the age of 84[109].

Henry Gardner appeared in a directory for 1809 as "watchmaker and dentist, successor to J. Rider"[110] and although the "dentist" part of this description disappeared, he was not involved in a single trade but throughout his working life had a number of strings to his bow. By 1819 he was described as "watch and clockmaker, silversmith and jeweller, optical and mathematical instrument maker"[111]; it is unlikely that he made the artefacts of all of these trades simultaneously, but could provide articles from each, and perhaps carry out repairs. Gardner's partnership with Job Rider lasted from 1805 to 1807, and he was then in partnership with Robert Neill between 1809 and 1818[112]. Neill, who is described as a "watch and clockmaker, jeweller, silversmith, optical and mathematical instrument maker"[113] in 1819, had apparently been an apprentice of Job Rider[114]. Before and after the partnership with Gardner, Neill conducted business in his own name alone, with his stock as varied as the description in the street directory; he also took over the business of another successful Ulster clockmaker who dabbled in astronomy on his retirement in 1816: John Knox[115]. Knox apparently wrote a number of scientific papers, including his own observations of the planet Venus, and was a founder member of the Belfast Literary Society[116]. He, too, sold instruments, although he described himself as a "watchmaker and jeweller" in an advertisement of 1809 which also announced "OPTICAL INSTRUMENTS, Telescopes, Reading Glasses, Spectacles . . . Mathematical Drawing Instruments, Clock and Watch makers' Tools, Materials &c"[117].

Little is known about the size of the workshops of any of these businesses, although occasionally at the bottom of a newspaper advertisement there is a request for workmen or apprentices[118]. It does appear, however, that the lines in clocks and watches and silverware were promoted more vigorously than the stocks of "Telescopes, Barometers, Brewers and Other Thermometers, Hydrometers, Quadrants and Ship's Glasses, Illuminators, Sun Dials, Electrifying Machines, Magic Lanterns, Circumferentors, Theodolites, Microscopes"[119] as offered at the end of an advertisement for a large range of goods from Gardner & Neill in 1810. This trend seems to have continued for much of the nineteenth century, with what little instrument making there was giving way to increased retailing.

Clockmaking remained a strength with a number of firms. Henry Gardner won a prize from the Board of Longitude for the construction of a chronometer in 1827, and used the occasion to publish a letter of congratulations from Thomas Romney Robinson, the Astronomer of Armagh, to advertise his expertise[120]. That same year, as if to emphasise the demise of one of his other skills, he was appointed Belfast agent for John Kibble, a London wholesaler and mathematical instrument maker to the Royal Board of Works, to sell drawing instruments to "Engineers, Masters of Schools, Drawing masters and others"[121]. Robert Neill's firm was carried on by other members of the Neill family well into the twentieth century, by which time Sharman D. Neill advertised "Turret Clocks a speciality"[122]. Just before the outbreak of the First World War the firm proudly announced that "Mr Sharman D. Neill has the honour to be Senior Warder of the Goldsmiths' Hall, Dublin"[123] – a far cry from the contemporary activities of the most prominent instrument maker in Dublin, Sir Howard Grubb.

The pattern of trade throughout the nineteenth century in Belfast appears to be one which, never very securely centred on instrument making or retailing, moved away from the tenuous beginnings of an embryo trade just when the population and industry of the area might have helped it to develop. The critical timing of Belfast's growth coincided with a period when the craft basis of the industry modernised and production moved into factories. However, as Hackmann has shown, even there the trade clung to its pre-industrial patterns, with outworkers merely moving under the one roof, and specialisms remaining[124]. Yet it was the market for instruments which does not seem to have developed in Belfast; at the industrial exhibitions held in Ireland in various places at various times throughout the nineteenth century, it is noticeable that Belfast remained unrepresented by local makers of scientific instruments. By 1895, when such an event was held in Belfast itself, contemporary advertisements clearly show that jewellery and watches were being promoted over instruments[125]. Similarly, a book originally published in 1891 to promote Ireland's industries mentioned only three Belfast businesses which might be classified as instrument makers. In all three accounts, it is apparent that this was not a main preoccupation, and that only some repair work, as opposed to actual manufacture, was undertaken[126].

Another indicator is advertisements in street directories, which generally echo this pattern. Only the relatively short-lived "Electrical and Optical Warehouse" of "J.A. Cumine, Manufacturer" appears to have produced at least some of the items that were sold. "Galvanic batteries, of every description, made to order" he proclaimed. "Air pumps, Magic Lanterns, Telescopes, Microscopes, and Mathematical Instruments, cleaned and repaired"[127]. That Cumine did indeed have the skills of the instrument maker is confirmed by the apparatus which Thomas Andrews (1813-1885) used to investigate the behaviour of carbon dioxide under a variety of pressures and temperatures[128]. The apparatus consists of two identical glass tubes. One contained carbon dioxide, the other, air. Pressure and temperature were controlled by a water bath surrounding the glass tubes. Pressure was measured by studying the behaviour of the air tube. The apparatus revealed the existence of a critical temperature, above which it was impossible to liquefy the gas. Andrews was the first Professor of Chemistry at Queen's College, Belfast, holding the chair from 1849-1879: it was natural that for his instrument making he should turn to a local man. The apparatus is signed "J.A. Cumine, 57 Arthur Street, Belfast". But for the most part, the established pattern – instruments brought in from outside and retailed – remained unbroken into the twentieth century.

Instrument making and retailing elsewhere in Ireland

Outside the main centres of population there was in general very little instrument making or trading. Lacking industry, Ireland's essentially rural hinterland did not need instruments, except in the specialised case of surveying. The same could be said for much of Scotland, and indeed the English provinces. For instance, Andrews quotes the example of the lack of instruments, and their makers, in eighteenth century Waterford[129]:

> Waterford 29 June 1724
> My Lord,
> I received from Col. Kennedy the petition of Alex Alcock, clerk, which was presented to your Lordship for the building of a barrack for Company of Foot in the old fort att Passage . . .
>
> If I could have gott any surveying instruments here, I would have made a plan of the place which would have explained itt much better to your Lordship than I can doe by writing
> I am &c.,
> JA: Kennedy.

Not until 1809 does Westropp list William Boncher, a mathematical instrument maker at Waterford, whose work is unknown. It was another 30 years before Abraham Prince, optician and mathematical instrument maker at Waterford made an appearance[130]; a single instrument is known with his signature, and may well have been a retailed item[131].

At Limerick, the pattern is similar; R. Wallace is known from a barometer and a level[132], but does not appear in the occasional street directories. Another name to be found on instruments from this locality is that of Charles Harrison, who appears to have been active during the mid-eighteenth century[133]. Jonathan Mason moved his business from Dublin to Limerick, when he married a local girl, and continued it until his death in 1849[134]. His son, James Stewart Mason, was a jeweller and goldsmith, working between at least 1843 and 1850; this gives some idea of how optical businesses did not always continue from father to son. James Mason married Cecilia, daughter of Lewis Pericho of Cork, carver and gilder[135], possibly an Italian immigrant.

The Italians who moved around the country put their names on instruments which are really domestic rather than scientific: barometers and thermometers. However, they often described themselves as carvers and gilders, or looking glass manufacturers, and it is difficult to tell whether an individual put an instrument together, or if he contributed the glassblowing element to its construction. Joseph Riboldi, carver and gilder, was trading in Limerick in 1824 and a number of barometers with his signature have been noted[136]. Westropp notes J. Castelli at Waterford in 1824[137]; Italian craftsmen had migrated across the Continent of Europe during the eighteenth century for economic reasons, and had found ready markets for their wares in France, the Netherlands and England. Goodison discusses how difficult it is to trace individuals, partly because they do not appear in documentary sources, but also because they were often still on the move[138]. In different localities names crop up which may be the same person, or the same family. There were, for instance, barometer makers named Malacrida in London, Manchester and Dublin; a barometer signed "Malacrida & Co Dublin & in London" is proof that at least two branches of the firm were run by the same people[139]. The size and variety of the London market attracted a large immigrant population of glassblowers[140], and the size of its port enabled easy opportunity to travel to Ireland, with which there was a considerable trade.

References and Notes for Chapter Five

1. Herbert Webb Gillman, *Index to the marriage licence bonds of the Diocese of Cork and Ross, Ireland, for the years 1623 to 1750*, Cork, 1896, 130.
2. The earliest edition we have located is the eighth edition, revised by Daniel Voster, and published in Dublin in 1764.
3. Gillman, *op. cit.*, 130 (ref. 1).
4. *Ibid.*
5. *Cork Evening Post*, 12 May 1760. Reference from Rosemary ffoliott's index in University College, Cork. The Science Museum, London, has a slide rule signed "Daniel Voster Corke Fecit 1742", inventory no. 1907-108. A sundial, signed "D Voster Corke Fe 1751" is in a private collection.
6. *Dublin Gazette*, 17 May 1760. Reference from Rosemary ffoliott's index.
7. *Cork Evening Post*, 16 June 1760.
8. *Corke Journal*, 8 September 1760. Reference from Rosemary ffoliott's index.
9. *Cork Evening Post*, 6 October 1763.
10. Inventory no. S.301.
11. *The Harp of Erin*, 10 March 1798.
12. William O'Sullivan, *The economic history of Cork from the earliest times to the Act of Union*, Cork and London, 1937, 223.
13. *Ibid.*, 129.
14. S.F. Pettit, *This city of Cork, 1700-1900*, Cork, 1977, 82-97.
15. O'Sullivan, *op. cit.*, 228-241 (ref. 12).
16. *Cork Evening Post*, 9 December 1773.
17. *Ibid.*, 19 August 1773 and the *Hibernian Chronicle*, 16 August 1773.
18. D.J. Bryden, Balthazar Knie, a provincial barometer maker, *Connoisseur*, **129**, 1972, 172-175; Nicholas Goodison, *English barometers 1680-1860*, 2nd edn., Woodbridge, 1977, 170-175. In 1798, Knie gave his age as 60 (i.e. born c.1738), born in Bohemia, and principal place of residence before arrival in Scotland "Cork, Ireland", entering at Irvine on the Ayrshire coast, Edinburgh City Archives, Register of Aliens 1798 [1803]. We are grateful to Dr A.D.C. Simpson for this reference.
19. John Smith, An old Edinburgh barometer maker, *The Weekly Scotsman*, **10**, December 1898, 7. In 1794, Knie declared he had been in Edinburgh for 19 years (i.e. arrived c.1775). Edinburgh City Archives, Declarations of Foreigners, 7th March 1794. We are grateful to Dr A.D.C. Simpson for this reference.

20. M.D. Westropp, "List of Dublin opticians and mathematical instrument makers", typescript at the Museum of the History of Science, Oxford.
21. Richard Lucas, *The Cork directory*, Cork, 1787, reprinted as The Cork directory for the year 1787, *Journal of the Cork Historical and Archaeological Society*, **72**, 1967, 135-157, and in *The Irish Genealogist*, **4**, 1969, 37-46. Alexander and Francis Fitton were probably related to William Fitton, whose work is represented by an octant signed "Made by William Fitton Corke for Thomas Hickman Esq 1773"; we are grateful to Dr Charles Mollan and Peter Delehar for this information.
22. *Ibid.*
23. M.V. Brewington, *Peabody Museum collection of navigation instruments*, Salem, Massachusetts, 1963, 14 and plate xi.
24. In a list of deaths: "yesterday suddenly Mr Francis Fitton of Paul St", *Cork Advertiser*, 17 September 1801. Reference from Rosemary ffoliott's index.
25. *Corke Journal*, 7 December 1753 and *Hibernian Chronicle*, 24 September 1781. Reference from Rosemary ffoliott's index.
26. The boy's name was George West. M.V. Conlon, Register of the Boys of St. Stephen's Hospital Cork, *Journal of the Cork Historical and Archaeological Society*, **62**, 1957, 41-55.
27. *Cork Evening Post*, 14 August 1783. Reference from Rosemary ffoliott's index.
28. D.J. Bryden, *Scottish scientific instrument-makers 1600-1900*, Edinburgh, 1972, 21-22.
29. "yesterday morning at his house in Bowling Green Street, in his 64th year, Mr Michael Fitzgerald, land surveyor", *Cork Constitution*, 22 September 1824. Reference from Rosemary ffoliott's index.
30. See Chapter Eight.
31. P. Eden (ed.), *Dictionary of land surveyors and local cartographers of Great Britain and Ireland 1550-1850*, Folkestone, 1975, 102.
32. Edwards and Savage, *Catalogue of books in the various branches of literature; stationery, maps, charts, patent medicines*, Cork, 1809, iv.
33. John Francis Maguire, *The industrial movement in Ireland, as illustrated by the National Exhibition of 1852*, Cork, 1853, 89.
34. *Ibid.*, 91.
35. Advertisement in Robert H. Laing, *Cork mercantile directory for 1863*, 27. Cobh was known as Queenstown from 1849-1922.
36. *Handbook to the harbour and City of Cork*, Cork, 1852, [67].
37. Laing, *op. cit.*, 27 (ref. 35).
38. Maguire, *op. cit.*, 90 (ref. 33).
39. Conlon, *op. cit.*, (ref. 26).
40. The dating is on the basis of Buckley's known career. However, Buckley has overwritten the name of a London maker on an octant in the National Museum of Ireland; information from Dr Charles Mollan.
41. Maguire, *op. cit.*, 89 (ref. 33).
42. [William?] Shaw, Report on the Department of Physical and Experimental Sciences, *Report of the Cork Cuverian Society for the Cultivation of the Sciences for the Session 1854-55*, Cork, 1855, 7-8.
43. E.G.R. Taylor, *The mathematical practitioners of Hanoverian England 1714-1840*, Cambridge, 1966, 356.
44. Sotheby's 21 July 1983 Lot 175. The signature is given as "F. Bennett Cork": almost certainly a misprint.
45. Sotheby's 20 September 1983 Lot 98.
46. Town Docks Museum, Hull.
47. These have been recorded by Dr Charles Mollan. The sextant is in the Department of Engineering, Trinity College, Dublin; the Sikes hydrometer in the National Museum of Ireland; the clinometer in University College, Galway; the other items are in University College, Cork.
48. National Maritime Museum.
49. Thomas Dix Hicks, *A syllabus of a course of lectures delivered in the year 1803*, Cork, 1803, [iv].
50. Margaret MacSweeney and Joseph Reilly, The Royal Cork Institution, *Journal of the Cork Historical and Archaeological Society*, **62**, 1957, 22-36; and *ibid.*, The last years of the Royal Cork Institution, *ibid.*, **63**, 1958, 23-25.
51. S.F. Pettit, The Royal Cork Institution: a reflection of the cultural life of a city, *ibid.*, **71**, 1976, 70-90, p. 79.
52. *Connor's Cork directory for 1812*, Cork, [1812].
53. *Connor's Cork directory for 1817*, Cork, [1817].
54. Ronan O'Rahilly, *A history of Cork Medical School*, Cork, 1949, Centenary Series no. 3, 24-26.
55. *Henry and Coughlan's general directory of Cork for 1867.*
56. *Slater's (late Pigot and Co.'s) Royal national commercial directory of Ireland*, Manchester, 1870.
57. *Wilkie's Cork city directory 1872.*
58. *Dublin, Cork and South of Ireland: a literary, social and commercial review*, London, 1892, 184-185.
59. Maguire, *op. cit.*, 89 (ref. 33).
60. *Dublin, Cork . . ., op. cit.*, 184-185 (ref. 58).
61. *Ibid.*, 185.
62. *General valuation of the rateable property in Ireland: Municipal Borough of Cork: Union of Cork: valuation of the several tenements*

comprising the Borough above named, Dublin, 1852, 1.

63. Goodison, *op. cit.*, 332 (ref. 18).
64. *Guy's directory for 1907*, Cork, 1907, 232.
65. *Slater's Royal national commercial directory of Ireland*, Manchester, 1881.
66. Those listed as opticians in *Guy's directory for 1907* were:
    Belgian-Franco-American Optical Co., 12B Winthrop Street.
    Brennan, Francis, Jeweller, 83 Patrick Street.
    Barriscale, William, Jeweller, 5 Bridge Street.
    Cork Chemical and Drug Co., Chemists and Oil and Colour Merchants, 80-81 Patrick Street.
    Egan, W., & Sons, Jewellers, Watchmakers etc., 32 Patrick Street.
    Fielding, P.J., FIO FSMC FCS, Chemist-optician, 66 Patrick Street.
    Gilbert, John, Fine Art Warehouse, 120 Patrick Street.
    Hilser, J., & Sons, Watchmakers and Jewellers, 95 South Main Street.
    Mangan, James, Ltd., Watchmakers, Silversmiths and Opticians, 3 Patrick Street.
    O'Sullivan, W., & Sons, Jewellers, 29 Grand Parade.
67. Hunt & Son appear in the directories for 1895, but not for 1896, when their long-occupied premises at 109 George's Street were listed as being vacant.
68. Maguire, *op. cit.*, 89-90 (ref. 33).
69. Pettit, *op. cit.*, 173-174 (ref. 51).
70. For Maguire (1815-1872) see G. LeG. N[orgate], *Dictionary of national biography*, **35**, London, 1893, 330-331.
71. *Annual Journal of the Engineering Society, University College, Cork*, **2**, 1914, 32.
72. Inventory no. 970.
73. The earliest appearance is in Alexander Aldwell, *The City and County of Cork Post Office general directory 1844-5*, the latest, *Slater's* ..., (ref. 56). He is not in *Francis Guy's County and City of Cork directory for the years 1875 and 1876*.
74. *General valuation of rateable property in Ireland: County of Cork: Barony of Kinsale: primary valuation*, Dublin, 1852, 19.
75. Leslie Clarkson, The city and the country, in J.C. Beckett *et al.*, *Belfast: the making of the city*, Belfast, 1983, 159.
76. J.C. Beckett, Belfast to the end of the eighteenth century, in *ibid.*, 18.
77. W.A. Maguire, Lords and landlords – the Donegall Family, in *ibid.*, 30.
78. L.A. Clarkson, Population change and urbanisation in 1821-1911, in Liam Kennedy and Philip Ollerenshaw (eds.), *An economic history of Ulster 1820-1939*, Manchester, 1985, 138.
79. *Ibid.*, 137.
80. Clarkson, *op. cit.*, 161 (ref. 78).
81. John Butt, Belfast and Glasgow: connections and comparisons, 1790-1850, in T.M. Devine and David Dickson (eds), *Ireland and Scotland 1600-1850*, Edinburgh, 1983, 193-203.
82. For some discussion of the nature of the trade at this time, see R.G.W. Anderson, Were scientific instruments in the nineteenth century different? Some initial considerations., in P.R. de Clercq (ed.), *Nineteenth-century scientific instruments and their makers*, Leiden, 1985, 1-12.
83. *Belfast News-Letter*, 10 April 1764. We are most grateful to Dr John Green for these references from the *Belfast News-Letter*, obtained from his computer index of the newspaper 1753-1790.
84. *Ibid.*, 12-16 December 1783.
85. *Ibid.*, 6-9 February 1776.
86. *Ibid.*, 12 September 1760.
87. *Ibid.*, 14-18 October 1774.
88. Exhibited in B.J. Mackay, *The Huguenots & Ulster 1685-1985*, Lisburn, 1985, exhibit no. 42(i).
89. John Hewitt, "The Northern Athens" and after, in Clarkson, *op. cit.*, 73 (ref. 75).
90. A selection of twenty-four papers published by the Society in 1981, N. McNeilly (ed.), *Selection from 150 years of Proceedings*, Belfast, 1981, contains four papers relating to instrumental science.
91. John Anderson, *History of the Belfast Library and Society for Promoting Knowledge, commonly known as the Linen Hall Library, chiefly taken from the minutes of the Society and published in connection with the centenary celebrations in 1888*, Belfast, 1888, 15. We are grateful to J.R.R. Adams for this reference.
92. *Ibid.*, 27.
93. *Ibid.*, 30.
94. *Belfast News-Letter*, 8 December 1767.
95. *Ibid.*, 12-16 August 1774. As with the Scotsman Alexander Wilson, who claimed to have invented philosophical bubbles in the mid-1750s, this was an adaptation of a principle known and used in the 1660s by the Accademia del Cimento: see Bryden, *op. cit.*, 34-35, n.181 (ref. 28).
96. *Ibid.*, 27-30 April 1779. A wheel barometer, signed "J. & P. Noseda, Walsall" was offered for sale at Phillips 2 December 1987 Lot 6; no trace of either Noseda was found in local directories between 1767 and 1834, but one Francis Noseda, glass-warehouse owner, Stanton Yard, Rushall Street, Walsall, was recorded in the census return for 1811. The International Genealogical Index reveals four baptisms of children of Francis and Ann Noseda at the Roman Catholic Chapel of SS Peter and Paul,

Wolverhampton between 1816 and 1826. Our thanks to the Local Studies Department, Birmingham Reference Library and Walsall Local History Centre, for this information. Whether Francis Noseda was related to John Noseda remains unresolved. A wheel barometer with paper scales signed "John Noseda fecit" was offered for sale by Sotheby's Sussex 19 June 1986 Lot 1958.

97. *Ibid.*, 20 March 1810. Bryden, *op. cit.*, 34, n.180 (ref. 28) discusses the specialist nature of the work of glassblowers.
98. Thomas Bradshaw, *Directory of Belfast and Lisburn*, Belfast, 1819.
99. Bryden, *op. cit.*, 46 (ref. 28) records Joseph Cappo at various Glasgow addresses between 1847 and 1853. A wheel barometer signed "JOSH. . CAPPO BELFAST" has been noted by Dr Charles Mollan.
100. Stick barometer recorded in an antique shop in Edinburgh, May 1987.
101. *Belfast News-Letter*, 19-22 March 1782.
102. Lucas, *op. cit.*, (ref. 21).
103. *Belfast News-Letter*, 19-22 March 1782.
104. S.D.T., Old County Armagh clockmakers, *Armagh Guardian*, 6 May 1938.
105. James Stuart, *Historical memoirs of the City of Armagh*, Belfast, 1819, 523, 525. Waugh subsequently moved to Dublin in 1805, where he continued clockmaking: *Antiquarian Horology*, **17** (1988), 377.
106. John S. Reid, A select clock, *Antiquarian Horology*, **13**, 1981, 45-50.
107. Quoted in J.J. Marshall, Ancient Ulster clockmakers, *Northern Whig*, January 1924 and *Belfast Telegraph*, 29 September 1930. Cuttings and manuscript notes in the Henry Collection, the Queen's University, Belfast; typescript in the City Library, Belfast.
108. Anderson, *op. cit.*, 27 (ref. 91).
109. Marshall, *op. cit.*, (ref. 107).
110. *Holden's trienniel directory . . .*, 4th edn., *for 1805, 1806, 1807*, vol. II, London, 1805.
111. Thomas Bradshaw, *Directory of Belfast and Lisburn*, Belfast, 1819.
112. Geraldine Fennell, *A list of Irish watch and clock makers*, Dublin, 1963, 15.
113. *[Joseph] Smyth's directory to Belfast and its vicinity . . . [for] 1820*, Belfast, 1819.
114. Colin Johnston Robb, Ulster's first watchmakers and jewellers, *Belfast Telegraph*, 6 February 1941.
115. Fennell, *op. cit.*, 22, 29 (ref. 112).
116. Discussed in Wilfred A. Seaby, James Wilson, clockmaker of Belfast, *Antiquarian Horology*, **14**, 1983, 152, and Jack London, The Clockmaker, *Belfast Telegraph*, 11 November 1964.
117. *Belfast News-Letter*, 15 October 1809, and repeated again 10, 17 and 24 November.
118. For example, *ibid.*, 5 January 1810 when Gardner & Neill required four clock and watchmakers; and 23 October 1827 when Henry Gardner states "Two Apprentices wanted".
119. *Ibid.*, 3 April 1810.
120. *Ibid.*, 9 January 1827.
121. *Ibid.*, 2 April 1827.
122. Description used in street directories from 1894. Also two trade catalogues, one dating from 1904, the other slightly later, make it clear that Sharman D. Neill was more of a jeweller and clockmaker; the turret clocks get a special mention. The trade catalogues are in the Library of the Ulster Folk and Transport Museum, Cultra.
123. *Ibid.*
124. W.D. Hackmann, The nineteenth-century trade in natural philosophy instruments in Britain, in de Clercq, *op. cit.*, 53-91 (ref. 82).
125. *Art and Industrial Exhibition Belfast 1895: official catalogue*, Belfast, [1895]. For instance, four pages of advertisements for Gibson & Co. Ltd make it clear that instruments were not a priority: Rankin Bros. were primarily jewellers; Steel & Sons provided such a vast variety of watchmakers' tools, "tools for scientific instrument makers, dentists, machinists, amateur mechanics etc." that they must have been retailers; Alfred Ansell promoted watches and jewellery.
126. *The Industries of Ireland, part 1 Belfast and the north . . .*, London, 1891: reprinted as *Industries of the North: one hundred years ago. Industrial and commercial life in the north of Ireland, 1888-91*, Belfast, 1986. This discusses Steel & Sons Ltd, Gibson & Co., and Rankin Bros.
127. *[John] Henderson's Belfast directory and northern repository*, Belfast, 1852, advertisements, 32.
128. Science Museum inventory no. 1876-1047. Described in J.A. Chaldecott, *Handlist of the collection relating to heat and cold*, Part 2, London, 1954, 21.
129. J.H. Andrews, *Plantation acres: an historical study of the Irish land surveyor*, Belfast, 1985, 250, n.92; S.S. de Val, Some eighteenth century petitions, *The Past: the organ of the Ui Cinsealaigh Historical Society (Wexford)*, no. 9, 1972, 10.
130. Westropp, *op. cit.*, (ref. 20).
131. A brass three draw telescope, signed by A. Prince, Waterford, was offered for sale at Christie's South Kensington 24 September 1976 Lot 25.
132. The barometer is at St Patrick's College, Maynooth, while the level is in the Department of Engineering, University College, Galway.
133. A portable sundial, signed "Harrison, Limerick"

was offered for sale at Christie's 5 July 1971 Lot 40; Dr Charles Mollan has noted a horizontal dial in Limerick. Fennell, *op. cit.*, 18 (ref. 108) gives Charles Harrison, clockmaker, watchmaker, maker of gold weights and scales as active between 1766 and 1786, and notes that Westropp had recorded a sundial with his signature.

134. [Thomas H. Mason & Sons Ltd], *The Mason family business: a brief history 1780-1980*, n.p., n.d. [Dublin, 1980], 4.
135. R. ffolliott, Biographical notes on some Cork clock and watchmakers, *Cork Historical and Archaeological Society Journal*, **69**, 1964, 49.
136. One was offered for sale at Phillips 8 May 1985 Lot 1.
137. Westropp, *op. cit.*, (ref. 20).
138. N. Goodison, The foreign origins of domestic barometers 1800-1860, *Connoisseur*, **170**, 1969, 76-83.
139. Offered for sale at Sotheby's 14 May 1987 Lot 226.
140. L.G. Rule, *The flame of the lamp*, London, 1967.

CHAPTER SIX

THOMAS AND HOWARD GRUBB

J.E. Burnett

Introduction

The Earl of Rosse's reflecting telescope with a six-foot mirror is perhaps the most famous scientific instrument which has been either constructed or used in Ireland. The present chapter gives only a brief account of it, since a monograph is available on the astronomical activity of the Rosse family[1], and the telescope has been discussed elsewhere by H.C. King[2]. Since the remainder of this book is concerned with the trade in instruments, and only incidentally notes the products of the amateur craftsmen, we may record that the Earl of Rosse did not make mirrors solely for his own telescopes: in 1843 he presented a 15-inch mirror of his own manufacture to Armagh Observatory[3].

Following a brief discussion of the work of another maker of specula for reflecting telescopes, C.E. Burton, we then describe the work of the firm of Thomas and Howard Grubb. Here another word of explanation is called for. The Grubbs have an important place in the history of science, and fully deserve to be the subjects of a book of their own. It is not our intention to write that book. Instead, we have outlined their history, and attempted to characterise their firm, and in particular to relate it to the history of science in Ireland.

The Earl of Rosse and the Leviathan of Parsonstown

> Welcome to thy new existence,
> Child of Intellect and Might!
> Welcome to thy wide Dominions,
> Lord of Ether and of Light![4]

The brothers William and Laurence Parsons, the earliest members of the Parsons family to live in Ireland, arrived in the country about 1590[5]. They were typical of the men whom Queen Elizabeth sent to colonise the island "beyond the Pale" – able, ambitious and loyal servants of the English crown. Laurence became Attorney-General for Munster, and among William's offices was that of Surveyor-General of Ireland – a post which we have already mentioned in connection with scientific instruments (see Chapters Two and Three). It was Laurence Parsons who settled in a village known as Ely O'Carroll, subsequently Parsonstown, and now Birr. Geographically, it was in the middle of Ireland; scientifically, it was in the middle of nowhere.

When William Parsons (1800-1867), third Earl of Rosse, was born, his family had been in Ireland for two centuries. Under the name of Lord Oxmantown he was educated at Trinity College, Dublin and Magdalen College, Oxford. A brilliant mathematician, he graduated from Oxford in 1827, and almost immediately began research into ways in which large reflecting

6.1. The "Great Birr Telescope", contemporary photograph. Royal Dublin Society.

telescopes might be improved[6]. In 1828 he produced his first machine for grinding and polishing mirrors, and experiments continued until 1845, when the six-foot reflector, "The Leviathan of Parsonstown" (*Fig.* 6.1), was brought into use. During the earlier part of his life, the Earl of Rosse was more concerned with the telescope than with astronomy: astronomy gave him a goal, but practical mechanics was his real enthusiasm. Later, his interest in mechanics was expressed in a series of letters on the design of ironclad warships[7], and in 1857 he was President of the Mechanical Science section of the British Association for the Advancement of Science.

Though James Short of London had made some fairly large reflecting telescopes in the middle of the eighteenth century, little work of scientific significance had been done with them. Parisian instrument makers, such as Passement, had sold instruments of a similar size. The possibilities of the reflector were spectacularly indicated in 1781 by William Herschel (1738-1822), who discovered Georgium Sidus – now familiar as the planet Uranus – with an instrument the mirror of which was 6.2 inches in diameter[8]. In October 1783 Herschel completed the 18.8-inch reflector which he employed in most of his researches into the "construction of the heavens"[9]. Three years later he started to use a 48-inch reflector with a 40-foot focal length, but its mounting was unwieldy and it was not used after 1815[10]. Rosse's work led on from Herschel's: Rosse, too, wanted to build large instruments[11]. This may be seen as being distinct from the continental practice of constructing comparatively large refractors. Joseph Fraunhofer (1787-1836) had developed techniques for producing a better quality optical glass and grinding object glasses to a high degree of optical perfection, and other instrument makers came to know of his methods. The 14-inch achromatic lens which E.J. Cooper purchased from Cauchoix of Paris in the early 1830s was one of the largest lenses in the world – and significantly smaller than Herschel's reflectors. Cooper had his lens mounted at Markree Castle, Co. Sligo by Thomas Grubb in 1834.

* * *

Skill and Strength in potent union,
Drew thee from the depths of Earth;
Art inventive, conquering Nature,
Gives thee now thy fiery birth[12].

The Earl of Rosse's success in making large metal mirrors was based on careful metallurgical tests. Herschel had used an alloy of 71% copper and 29% tin by weight – a composition which had originally been recognised by Samuel Molyneux (1689-1728), son of William Molyneux of the Dublin Philosophical Society[13]. Herschel had tried using alloys which had a higher proportion of tin, and thus gave a brighter polished surface, but this metal was so brittle that mirrors cast from it shattered as they cooled. Rosse discovered that a brighter alloy, about 31.5% tin, could be used if the mirror were cooled slowly in an annealing furnace[14].

At this period there was considerable interest in the use of novel materials in optical instruments. In attempts to make large object glasses, experiments were carried out with fluid lenses – glass cells containing a liquid, typically hydrochloric acid or carbon bisulphide[15]. Rosse himself tried to produce an effective fluid lens[16]. Advances in the understanding of polarised light brought calcite and selenite into prominence in the 1820s. The Nicol prism was invented in 1829[17]; investigations were undertaken into the practicality of using jewels or precious minerals with a high refractive index for microscope objectives, but with only limited success[18].

It was difficult to cast a speculum, but more difficult to grind and polish one[19]. From the beginning of his experiments, the Earl of Rosse used a steam-driven machine, first to grind a spherical figure, then to polish it to a parabola[20]. The polishing was done by a block of pitch attached to an iron backing. A small technical improvement of particular significance was the cutting of deep grooves in the pitch, and in the iron which held it[21]. The grooves carried away both excess abrasive, and also the abraded material, and led to the mirror having a more accurate figure. Quite extraordinary care was taken to eliminate any factor which might have disturbed the accuracy of the result. For example, during the polishing of the mirrors for the 36-inch and 72-inch telescopes, the whole polishing machine and mirror were immersed in a water bath maintained at a constant 55°F[22]. It is hardly surprising that the mirrors thus produced were of a very high quality.

* * *

Genius, led by wary Science,
Shall unveil thy face to light,
Like the spirit that hath formed thee,
True and stainless, firm and bright[23].

The 36-inch reflector which Rosse completed in 1839 (*Fig.* 6.2) was the largest telescope in operation in the world. He immediately began to lay plans for an instrument twice its size. The great reflector was brought into use in 1845. It was a stupendous achievement, more akin to the contemporary engineering works of men such as Isambard Kingdom Brunel, and Robert Stephenson, than to any scientific instruments. The mirrors alone weighed 4 tons, and the telescope tube was slung between two parallel walls 72 feet long and 56 feet high. The tube itself was 58 feet long. It has been said that the cost was £20,000[24].

The year 1845 might have been remembered as one of modest progress in Ireland. The railway system, lagging a little behind that of England, was beginning to develop, and Kingsbridge (now Heuston) station was being built in Dublin, to serve the south and west of the country. The Museum of Irish Industry opened, to forward technical instruction. And the great telescope began work at Parsonstown. All these events, however, are merely trivial when compared with the failure of the potato crop. Famine followed; the government in Westminster was unable to understand that death by starvation was common at a distance of two days' journeying. In the

6.2. Contemporary model of the 36-inch reflector at Birr Castle. Trustees of the National Museums of Scotland.

midst of the central plain of Ireland the Earl of Rosse had a different perspective, and abandoned astronomy for relief work[25]. The telescope was brought back into regular operation in 1848.

The scientific importance of the 6-foot reflector lay in its ability to gather more light than any other telescope in the world, and thus to detect the detailed structure of very faint astronomical objects. The discovery of the spiral structure of a galaxy created a sensation[26], but of equal importance was the revelation that some nebulae, hitherto thought to consist of a continuum of fluid material (the word "gas" was not used), were in fact aggregates of immense numbers of stars[27]. This had a cosmological implication – that some nebulae at least were far further from the Earth than had been supposed, and this implied that the universe was considerably larger than had been believed.

The Leviathan of Parsonstown was dismantled after some years of disuse in 1908[28]. One of the two 72-inch mirrors is now in the Science Museum, London[29].

* * *

On then to thy heavenly office!
Which the best of Earth acclaim,
Honour to the Master's nation!
Praise undying to his Name![30]

The poem which has been quoted was the work of one of the Earl of Rosse's closest scientific friends, the Reverend Thomas Romney Robinson. Robinson (1792-1882), was Director of Armagh Observatory for the last 59 years of his long life. A frequent visitor at Parsonstown, Robinson can be seen as a member of the social circle of astronomers which was centred on the Earl of Rosse. Sir William Rowan Hamilton, the great mathematician, who shared many philosophical and poetic enthusiasms with Robinson, was in the same circle. So, too, were the various young men whom the Earl employed from time to time as tutors and observers. Several of them later attained distinction in Irish science, and one of them emulated the Earl's practical interests – Charles Burton.

Charles Burton

Charles Edward Burton (1846-1882) was employed at various times by the Earl of Rosse, and by Dunsink Observatory, and also by the Royal Observatory, Greenwich. He was with the last named of these institutions on the expedition to observe the transit of Venus of 1874 from Rodriguez Island, near Mauritius[31]. His father was the rector of Loughlinstown, eight miles south-east of Dublin, and it was whilst he was at Loughlinstown that Burton figured telescope mirrors. He learnt the art of speculum grinding from the Earl of Rosse, although Burton made not metal specula but silvered glass mirrors. In size they ranged from 6- to 15-inches diameter, and an obituarist described the best of his mirrors as being "unsurpassed"[32]. For his own observations he used 8- and 12-inch mirrors, with excellent results[33], and he supplied mirrors to other astronomers. He can be compared with George With, Master of the Blue Coat School, Hereford, Rev. H. Cooper Key and J.T. Bird, amateur astronomers who took up mirror making in the decade or so following the introduction in 1856 of silvered specula by Carl August von Steinheil of Munich and Léon Foucault of Paris[34].

Irish astronomers were notably interested in novel instrumentation in the last quarter of the nineteenth century. W.H.S. Monck, a Dublin lawyer, was involved in the first attempts to apply photoelectric photometry to the stars and planets[35]. The fourth Earl of Rosse's development of instruments to measure the surface temperature of the Moon are well-known[36]. A few years later another amateur astronomer, W.E. Wilson, employed an industrial optical pyrometer to measure the surface temperature of the Sun[37]. Burton's friend Wentworth Erck, another lawyer, obtained the 7½-inch refractor which had been made by the famous American instrument maker Alvan Clark in 1854: it was known to be of the very highest quality[38]. Burton and Erck made a comparison between the Clark object glass and a 9-inch mirror which was presumably made by Burton[39]. The reflector, they said, gave the better definition. Burton made about a dozen mirrors, some of which he supplied to other astronomers, such as George Johnstone Stoney[40], and probably also Erck.

Charles Burton also designed instruments, such as a ghost micrometer, jointly with Howard Grubb[41], and a binocular spectroscope[42]. This latter and rather bizarre instrument was built to observe the very faint zodiacal light: the binocular arrangement was used in the hope that two eyes would observe the low light intensities more easily. It was partly made by Spencer and partly by Grubb. The use of a chain to focus the twin telescopes suggests the work of Spencer, who used similar chains in other instruments, and Spencer's trade card is on the box. Grubb probably made no more than the prisms and the recording mechanism. Burton took it to Rodriguez, and was able to observe with it on ten nights before the cement holding the prisms in place dried out[43].

Burton suffered from ill-health throughout his life: this was the cause of his discontinuous employment. Aged only 35, he died suddenly in church at Castleknock – the nearest church to Dunsink Observatory – whilst the congregation was singing "Nearer My God to Thee".

Thomas and Howard Grubb

Of all the Dublin instrument makers, only one firm had an international standing – that of Thomas Grubb (1800-1878) and his son Howard Grubb (1844-1931). Thomas Grubb's great, great grandfather, John, landed at Waterford in 1656[44]. It was the time when Ireland was oppressed by a "Protector": Cromwell failed to take Waterford in 1649, but it fell to his son-in-law Henry Ireton in 1650. John Grubb (c.1620-1696), son of Ishmael Grubb of Ravensthorp, Northamptonshire, was an Anabaptist preacher, who became a Quaker after 20 years in Ireland. The family appear to have stayed in the southern counties of Ireland, as merchants and minor landowners, until Thomas Grubb was born at Kilkenny in the last year of the eighteenth century. He should not be confused with his third cousin, Thomas Grubb or Grubbe of Clonmel, iron oil and colour merchant and boat-builder, who made the River Suir navigable as far as Clogheen and Cahir[45].

Thomas Grubb began business as a precision mechanic about 1830 at Charlemont Bridge on the south side of Dublin[46]. We do not know where he received his training. He constructed an observatory of his own, equipped with a 9-inch telescope[47], and soon became involved in the construction of large telescopes, the first one of importance being a 15-inch reflector with an equatorial mounting, supplied to Armagh Observatory in 1835. The last instrument he built was also his largest, a 48-inch Cassegrain reflector with a metal speculum (1865-1868), produced with the aid of his son, for the Melbourne Observatory. It was constructed in temporary buildings on a site off the Rathmines Road (rather less than half a mile south west of Charlemont Bridge), where the new works was subsequently built.

Howard Grubb received a training in civil engineering at Trinity College, Dublin[48]. Offered the chance to join the Royal Engineers, he preferred to enter his father's works[49]. He joined his father in business in 1865 and took control of the firm in 1868, moving to Rathmines in 1875-1876. In 1918 the factory at Rathmines was closed, and Grubb's operations were moved to St Albans in England; seven years later Howard Grubb retired and sold his company to another Irishman – Sir Charles Parsons, inventor of the steam turbine. Parsons removed the factory from St Albans to Newcastle-upon-Tyne[50]. Howard Grubb's largest telescopes were a 27-inch refractor for Vienna, completed in 1881 and for some years the largest refractor in the world; and a 40-inch reflector for the Simeis Observatory in the Crimea, put into commission in 1928, two years after he had retired.

During the nineteenth century, there was an enormous extension in the range of research carried out by astronomers. Astronomy had traditionally been concerned with the positions of the stars and planets. During the millenia since Ancient Babylon practical problems in navigation and calendar reform had required closer and closer attention to positional astronomy. Very little evidence was available as to the physical nature of the heavenly bodies, the movements of which were so widely studied. Improvements in the optical quality of telescopes, and the construction of large instruments such as those of Herschel and the Earl of Rosse, brought a more attentive examination of everything from the Sun and Moon to the faintest nebula. The universe was found to be more diverse, more complex and more interesting than had been supposed. The discovery of the technique of spectroscopic analysis by Bunsen and Kirchhoff in 1859-1860 brought a new dimension to astronomy: it became possible to unravel the chemical elements which constituted each star. By the end of the century, photography had further extended the range of information available, since long exposures could reveal sources of light too faint for the human eye to perceive.

Telescopes developed in parallel with astronomy. Larger instruments were needed to detect fainter objects. Both visual and photographic work required improved mountings, and when photographic plates were exposed for several hours, extremely smooth telescope drives were also essential. It was in this rapidly changing science that the Grubbs found their place.

The Grubbs' status

It must be emphasised that the Grubbs' work was of the very highest quality. The contract for the Melbourne 48-inch reflector (*Figs* 6.3a & b) marked them out as the leading makers of large reflectors in the British Isles – if not the world. In retrospect, it is clear that by the year of its completion, 1868, the metal mirror was an outmoded technology: yet who with confidence would have made a 4-foot silvered glass mirror? The Earl of Rosse's experience with metal mirrors should have made it clear that it was essential to repolish them frequently. The Astronomer at Melbourne had to learn to do this himself. The equatorial mounting, however, was far in advance of Rosse's. The instrument, comparatively, was a failure. The large reflector was discredited, and Grubb was not involved again in the construction of this type of instrument until the 1890s, when he mounted several mirrors made by A.A. Common of Ealing, London. Only in 1925 did Howard Grubb complete another really large reflector on his own.

The Grubbs built many large refracting telescopes. In the first decade of the nineteenth century, the problems of making large glass blanks which were free from defects had been largely solved, though knowledge of these techniques spread slowly from Germany, first to France, and then to England. It thus became possible to build larger and longer telescopes[51] though it is not fully clear to what extent the move towards enormous instruments was encouraged by the needs of astronomers for greater light-gathering power, or by the ability of the instrument makers to create more precise optics with greater dimensions, or by the ready supply of wealthy men who wished to make a conspicuous contribution to astronomical research. A comparative study of the motivation of such men as the London surgeon Sir Henry Thompson, the American property speculator James Lick[52], and the American tramway magnate Charles T. Yerkes, is long overdue. There was intense international competition to satisfy this market. The leading firms, such as Alvan Clark of Cambridge, Massachusetts, Repsold of Hanover, and Steinheil of Munich, were magnificent scientific craftsmen – and the Grubbs could hold their place among them, as a list of their more important instruments indicates (see the appendix to this Chapter).

Hermann Carl Vogel (1841-1907), one of the most conspicuous figures in late nineteenth century astronomy, was Director of the astrophysical observatory at Potsdam for 25 years. He had the opportunity to use the great Vienna refractor (*Fig*. 6.4) not long after its completion, and wrote,

6.3.(a) The 48-inch Melbourne reflector at Grubb's works, Dublin 1868. Royal Astronomical Society.

6.3.(b) Melbourne Observatory, with the 48-inch reflector, and Melbourne in the background, c.1870. Royal Astronomical Society.

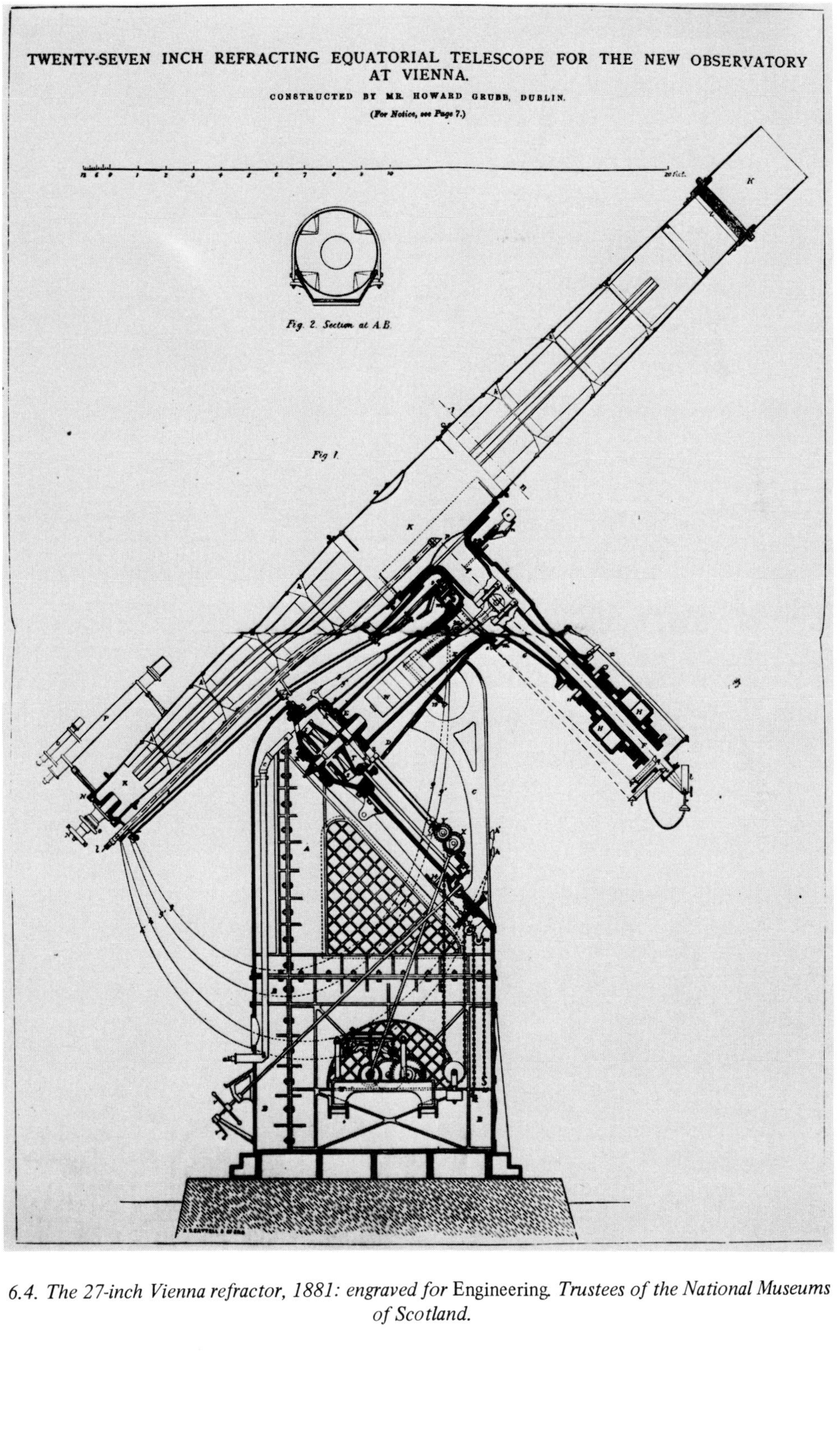

6.4. The 27-inch Vienna refractor, 1881: engraved for Engineering. *Trustees of the National Museums of Scotland.*

6.5. Cork 8-inch refractor at the Paris Exhibition, 1878. Royal Astronomical Society.

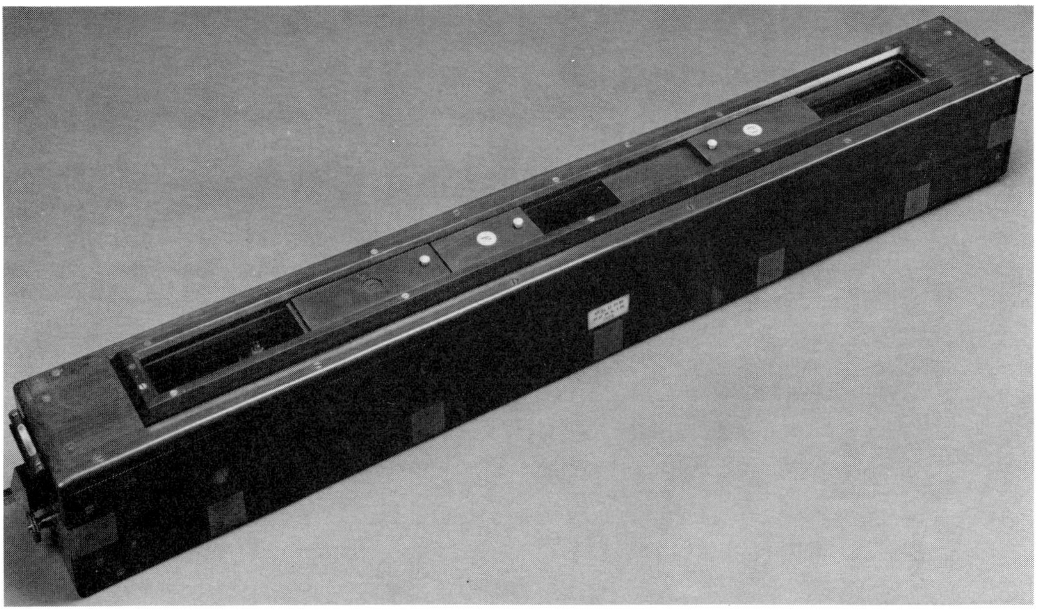

6.6. One of a set of baseline measures, made by Grubb for Lord Lindsay's expedition to Mauritius in 1874 to observe the transit of Venus across the sun's disk. Trustees of the National Museums of Scotland.

6.7. Six prism spectrometer for astronomical use, made by Grubb & Son, Dublin, for William Huggins, c.1870. Whipple Museum of the History of Science, Cambridge.

HOWARD GRUBB

(Late GRUBB & SON)

RATHMINES, DUBLIN,

Optical and General Mechanician to the Royal Society and the principal Public and Private Observatories of Great Britain and Colonies;

DESIGNER AND CONSTRUCTOR

Of the Spectroscopes and Spectroscopic Apparatus used in connection with the GREAT EQUATORIALS manufactured by him for

THE GOVERNMENT OF VICTORIA,
THE ROYAL SOCIETY,
THE EDINBURGH OBSERVATORY,

And Lord Lindsay's NEW OBSERVATORY at DUN. ECHT. ABERDEEN;

DESIGNER AND MANUFACTURER

OF

THE COMPOUND PRISMS

(First Introduced by T. Grubb, F.R.S. in the Spectroscope of the Great Melbourne Telescope),

ALSO

THE RECORDING SPECTROSCOPE AND THE AUTOMATIC SPECTROSCOPE

(First used by W. Huggins, Esq., LL.D., F.R.S.).

NOW IN HANDS:

Large Batteries of Spectroscopes for use with the

NEW 15-in. ACHROMATIC

Just finished for Lord Lindsay, and for the

NEW 12-in. ACHROMATIC

In hands for Oxford University. Also

Spectroscopes on a New Improved Automatic principle.
Liquid Prism with Telescopes 8-in. Aperture, &c. &c.

Price Lists and further information on application. May, 1873

6.8. Advertisement by Howard Grubb, 1873. Trustees of the National Museums of Scotland.

> In the spring months [of 1883], when there were several consecutive days of exquisite clear weather, I got the impression that the object glass was rather good, but that the images, as regards sharpness, were not to be compared with those of middle-sized instruments, and on leaving Vienna I had formed the opinion that the difficulties of producing so large Objectives had not been quite surmounted . . . I have acknowledged that the Vienna Objective, as regards the precision of the image, leaves nothing to be desired, and that it was only from want of taking the state of the air properly into account that I had formed my former opinion . . . The images of bright stars were of perfect regularity, and the central part of the diffraction disc was so remarkably small that it may be expected that the instrument would also be very suited for observing double stars[53].

At the end of a description of the 12¼-inch refractor at Oxford, the Savilian Professor wrote, "I cannot refrain from recording my admiration of the admirable manner in which Mr Grubb has much more than fulfilled his contract . . . English[!] opticians have at length overtaken their eminent brother artists on the Continent . . ."[54].

The Grubbs' work was performed on a grand scale. Between 1890 and 1914, when their production of large instruments was at its peak, they produced over 90 lenses for refractors with diameters ranging from 5 to 28 inches[55]. During this period they also made smaller instruments, spectroscopes of all sizes, military and naval optical equipment, as well as carrying out a variety of miscellaneous work, such as refiguring for A.A. Common the mirror of the 36-inch telescope which had been made by George Calver of Chelmsford: Common subsequently sold it to Edward Crossley, a carpet manufacturer in Halifax, who in 1895 presented it to the Lick Observatory[56]. In his 1908 catalogue Grubb listed 24 observatories which he had supplied with domes[57]. His domes were not trivial structures: the prices specified in the 1888 catalogue included £200 for an 18-foot diameter dome, covered with papier-maché, and £1,200 for one of 36-foot diameter, covered with iron plates[58].

It is particularly difficult to assess the number of smaller telescopes made by Howard Grubb. At the 1887 Manchester Exhibition, besides exhibiting the twin equatorial telescope designed for photographic work, he also had on display a 5-inch equatorial refractor, a 4-inch binocular telescope, his 3-inch student's equatorial with circles divided on paper, a 4-inch telescope mounted on the siderostatic principle, "as well as an excellent collection of spectroscopes, eye-pieces, object glasses in various stages of manufacture, &c"[59]. In 1888 he advertised "Small instruments (suitable for university students)". The cheapest model was a 1½-inch refractor on a transit mounting, which sold for £12[60]. He had appointed Watson & Sons of London as agents[61]. By 1908 the telescope with the lowest price was one with a 2½-inch object glass, and mahogany tripod with altazimuth head, which cost £15-10-0[62].

The Grubbs as precision mechanics

Although Thomas and Howard Grubb were optical instrument makers, their greatest skills were perhaps mechanical rather than optical. Thomas first made his name as the designer of machinery for engraving, printing and numbering banknotes, used by the Bank of Ireland until 1921, and he made dramatic improvements to the mechanics of the microscope (*Fig.* 6.9)[63] which were followed by other makers only some 50 years later. From time to time Howard wrote on Charles Babbage's method of mechanical notation[64], turret clocks[65], and the machinery for moving telescope domes[66]. He was also responsible for a hydraulic-powered rising floor which was designed for Lick Observatory, and later bought by a number of observatories, including the Radcliffe at Oxford and that in Santiago in Chile[67]. It enabled the observer to remain in a comfortable position without regard to the elevation of the telescope tube, or, as Grubb put it in one of his catalogues, it was "a solution of the troublesome question of the observing chair"[68].

It was in the mounting and driving of large telescopes that the Grubbs made their chief

6.9. Compound microscope, signed "Grubb, Dublin", c.1860. Museum of the History of Science, Oxford.

mark[69]. The 48-inch Melbourne reflector was their first large reflector to have an equatorial mounting, a notable advance on the primitive mountings used by the Earl of Rosse for his two largest telescopes[70]. The 27-inch Vienna refractor was given a mounting in which friction had been dramatically decreased from that in previous mountings. Whilst working on the Vienna refractor, Grubb introduced a new method of testing for spherical and chromatic aberration, by placing various lenses between the eyepiece and the objective. In explaining his method, he used a typically mechanical metaphor: "In fact the perfection of any correction in an objective means *the best balance between opposing aberrations*, and (just as in all cases of ascertaining balances) it is difficult to determine the neutral point unless there be the power of trying on both sides"[71].

Howard Grubb's interest in unorthodox telescope mountings also illustrates his mechanical enthusiasms. In the 1880s he built what he called a "siderostatic telescope" for solar observations[72]. Later, he built a 12½-inch polar coudé instrument which was installed in Cambridge[73]. He designed another arrangement which was proposed for a telescope for Armagh Observatory[74]. This was based on a "dialytic" telescope – one in which achromatism is achieved not by a doublet, but by a large crown glass objective and a flint (or flint-crown) secondary lens with negative power, which both gave a chromatic correction and a narrow cone of rays which would require – if the telescope were to be a coudé – only a small flat. In this form, Grubb believed that the instrument would be comparatively cheap, as well as having the advantage of all coudé telescopes, that the observer could remain stationary while the object he was studying moved. Grubb emphasised this point by publishing a lithograph in which the observer was seated in a comfortable armchair, complete with tassels, beside a blazing fire.

In addition to their concern to develop better telescope mountings, the Grubbs were also involved in improvements in the figuring and polishing of lenses and mirrors. Grinding and polishing machines are pieces of mechanical, rather than optical, apparatus. The first machine for grinding specula was made by William Herschel; the machines of the Earl of Rosse and William Lassell were improvements on it. It is not clear how the machines used by the Grubbs relate historically to the others. Thomas Grubb's first machine was presumably used from the time of the 1835 Armagh reflector until it was superseded by his second machine, which he employed for the Melbourne reflector[75]. There appears to be no evidence as to the type of machines which were used after 1870.

Hackmann has pointed out that there have been a number of instrument makers who were precision mechanics too[76]. James Watt and John Smeaton are eighteenth century examples; in the nineteenth century Thomas Cooke of York, the Société Genevoise and R.W. Munro of London fulfilled similar roles. The case of Munro is strikingly similar to that of Grubb for, in addition to a range of instruments, particularly seismographs, he also devised and constructed machines for printing banknotes for the Bank of England and postal orders for the General Post Office[77].

The demise of the firm of Grubb on its purchase by Sir Charles Parsons indicated a significant change. One of the advantages of moving the factory to Newcastle-upon-Tyne was that there the heavy machining could be carried out in the parent heavy engineering works[78]. The manufacture of large telescopes – in Britain at least – could not sustain itself without outside aid.

The Rathmines works[79]

The first buildings in Rathmines Road were temporary sheds used during the construction of the Melbourne reflector. They were erected in 1865, and were largely replaced by more durable buildings in 1874. The largest workshop was about 70 feet square, and beside it was a 12-sided room with considerable headroom, which had been built to contain the Vienna instrument whilst it was being assembled. Howard Grubb made all the parts of his instruments, starting from rough castings or glass blanks, so further accommodation was provided for a smithy, model makers, dividing engines, a machine shop, and an optical department. Between 35 and 40 men were employed in 1888.

The works contained much specialist equipment, most of it made by the Grubbs themselves. There was a small screw-cutting lathe for cutting micrometer screws, usually with 50 or 100 threads per inch. The crucial part of this machine – the lead screw – had been cut about 1850 by Bryan Donkin of London. Thomas Grubb built the early pantographic copying machine which was used to engrave the numbering on divided circles. This machine, which dated from 1847, was one of 30 made for the government. Nervous of the unrest which was to culminate in the rising of 1848, Parliament in London had passed an Act requiring that every firearm in Ireland should be registered, and should have a registration number engraved on it. The number of human engravers available was inadequate, so Grubb devised a machine which could be used by the constabulary. It is evident that quite apart from the practical skills Howard Grubb was taught by his father, he inherited unusual and effective plant.

Working conditions were at times appalling. During the casting of the metal mirrors (when the roof of the works was burnt off) the workmen wore aprons and gauntlets of thick felt, and calico hoods soaked in alum. The heat was so intense that the moment the casting was completed, Howard Grubb walked into the outside air and fainted. At other times the men had to travel in mid-winter to observatories, to erect and finish instruments, working at night by lamplight[80].

The Grubbs also made use of scientific instruments in the process of manufacture. For example, when glass blanks arrived at Rathmines Road a polariscope was employed to check whether they had been correctly annealed. In the stages before a lens or mirror was ready for optical testing, its curvature was checked with a spherometer which could measure distances as small as a fifty-thousandth of an inch. Howard Grubb's method of cutting the teeth in the drives of the large equatorial instruments was to first divide the wheel on a dividing engine, then to cut the teeth individually on a cutting machine, setting each tooth with the aid of a microscope.

The production of an optical surface on a large piece of glass was carried out in several stages. In the 1880s the costs of removing a gramme of glass in each of these four stages was as follows:

	£	s	d
Rough grinding	0	0	1
Fine grinding	0	0	7½
Polishing	0	10	0
Figuring	48	0	0

Rough grinding was neither more nor less than its name suggests. Fine grinding involved the use of ribbed tools of brass or iron, of the type invented by the Earl of Rosse, and finer and finer abrading powders. Lenses up to about six inches in diameter were ground by hand, but larger lenses required such physical effort that they were ground on a machine. Polishing was the continuation of grinding, using a tool coated with hard pitch. Finally came the figuring of the lens, a process which required an excellent knowledge of practical optics in order to distinguish the many tiny faults which remained in the lens. During figuring, repeated optical testing was carried out. "A man must have great natural qualifications and a scientific knowledge of a high order to enable him to reach success in such a field", wrote *Engineering*. "In addition he must possess exceptional patience."

The Grubbs and Ireland

The Grubbs' work does not belong to the tradition of instrument making in Dublin. The

6.10. Grubb's stand at the St Louis Exhibition, 1904.

instruments they made have no antecedents among those made for the Irish market by local tradesmen. The market which the Grubbs served was first British and then world-wide. The only continent to which they did not sell a telescope was Antarctica, and they won a gold medal at the International Exhibition at Paris in 1878 and a Grand Prize at St Louis in 1904 (*Fig.* 6.10)[81]. The 27-inch refractor for Vienna is probably the crucial instrument in the development of the firm. Not only was it very large, but it also represented the firm's first penetration of the market for large telescopes beyond the British Empire. The telescope was ordered in 1875[82], and one suspects that the ill feeling after the war of 1866 between Prussia and Austro-Hungary may have discouraged the purchase of a large instrument from German manufacturers. The optical design of the Vienna telescope was carried out by Sir George Gabriel Stokes (1819-1903), Lucasian Professor of Mathematics at Cambridge, son of an Irish country parson and great-grandson of Gabriel Stokes, the Dublin instrument maker[83]. In addition, Stokes gave Thomas Grubb extensive advice on the problems of testing telescope mirrors[84], and another one of the Grubbs' largest object glasses, the 28-inch for Greenwich, completed in 1893, was also made to Stokes' designs[85].

The instrument making firms which the Grubbs resemble are those such as Utzschneider & Fraunhofer of Munich, Thomas Cooke of York, the Cambridge Scientific Instrument Company, the Société Genevoise, and two Glasgow companies – James White (later Kelvin & White Ltd) and the partnership of Barr & Stroud. These firms share an important characteristic. They all made highly specialised apparatus for which they found a world, rather than a local, market, and therefore they had no need to carry out their business in one of the traditional centres of instrument making such as London, Paris or Berlin, which had grown up because of their convenience for the traditional markets. A measure of Grubb's lack of contact with the Dublin market is seen in the instrument purchase book of the Royal College of Science for Ireland. Between 1868 and 1879 the College placed over 30 orders costing more than £1. Most went to

Yeates or Spencer of Dublin, or to Elliott Brothers of London. Only one, for a photographic lens, went to Grubb [86].

In addition, the Cambridge Scientific Instrument Company and the two Glasgow firms owed their particular location to the involvement of academic scientists in their work. Horace Darwin, youngest son of Charles Darwin, was the moving force behind the Cambridge Scientific Instrument Company, and had very close links with the Cavendish and Physiological laboratories in Cambridge[87]. The development of James White's firm of opticians into one of the world's leading manufacturers of electrical measuring instruments by the end of the nineteenth century was due to the involvement of William Thomson, Lord Kelvin, Professor of Natural Philosophy at Glasgow who had been born in Belfast[88]. Kelvin's colleague, Archibald Barr, Professor of Civil Engineering, was one of the inventors of a highly successful range-finder which was manufactured in Glasgow so that Barr could be on hand to improve it and to ensure the development of the firm through the invention of further military and naval instrumentation[89].

Though neither of the Grubbs held academic positions, they did, however, take part in Dublin academic life, and Howard was the first Irish instrument maker, if not the first in the British Isles, to have embarked on a formal university education, at Trinity College, Dublin. Both Thomas and Howard were members of the Royal Irish Academy, and Howard received its Cunningham Gold Medal in 1881. He also received the Boyle Medal of the Royal Dublin Society in 1912[90]. Thomas Grubb published two articles in the *Proceedings of the Royal Irish Academy* and three in the *Journal of the Royal Dublin Society*. Between 1870 and 1912 his son made another 33 contributions to the Royal Dublin Society's various periodicals. This was the period when the Society was playing a leading part in the promotion of science in Ireland[91]. Howard Grubb was Vice-President of the Royal Dublin Society for many years; he was effectively debarred from being elected President by the convention that the senior post should always be held by an aristocrat. He was, however, appointed a Governor. Howard Grubb was also a Scientific Adviser to the Commissioners of the Irish Lights from 1913, and held the honorary degree of Master of Engineering from Dublin University. Thomas Grubb was engineer to the Bank of Ireland from 1840 until his death. Both father and son were Fellows of the Royal Society of London.

Amongst the Grubbs' scientific friends, one of the most important was Thomas Romney Robinson, Director of Armagh Observatory. With funds provided by Lord John George Beresford, it was Robinson who gave Thomas Grubb his first major astronomical commission. Robinson was the inventor of the familiar cup anemometer. Robinson's daughter by his first wife married George Gabriel Stokes, and his second wife was the daughter of Richard Lovell Edgeworth, and the half-sister of Maria Edgeworth, the novelist. Thomas Grubb named his fourth son Romney Robinson (b.1845). The child died in infancy, but the link with the Robinsons was sufficiently strong for Howard to name his own third son Romney Robinson Grubb[92]. When he was over 80 years old, Robinson made a redetermination of the constants of the cup anemometer, and the experiments were carried out in Grubb's works[93]. Anglo-Irish society was small, and complex patterns of family and social connections were not uncommon. When we learn that Richard Lovell Edgeworth investigated the problems of banknote engraving[94], and that Thomas Grubb constructed machines for printing banknotes, we may wonder whether distant coincidence was not the product of a collective subconscious. The Grubbs were unusual among instrument makers, however, in belonging to a family of minor landowners.

One of the most important commissions in the history of the firm was not astronomical. The Professor of Natural and Experimental Philosophy at Trinity College, Humphrey Lloyd, designed in 1839 instruments for a network of some 40 magnetic observatories in the British colonies[95]. The set of magnetometers (a horizontal force magnetometer, a vertical force magnetometer and a declinometer) which was installed on St Helena, is now in the Science Museum, London[96]. Lloyd found it most convenient to have the instruments made in Dublin by Grubb, despite the fact that he had no suitable facilities for producing the necessary magnets, and his pieces of steel had to be sent to E.M. Clarke, an expatriate Irishman in London, who had a sufficiently large electromagnet[97]. This episode illustrates again the importance of the Dublin scientific establishment to the Grubbs. Between 1839 and 1843 Trinity College paid Grubb £146-16-0 for magnetic apparatus[98]. The College also bought magnetic equipment from Thomas Jones of London[99], from Robinson of London and Gambey of Paris[100].

In the same way, despite their later international status, they were still very much a local firm when Thomas Grubb was awarded the contract for the 1835 Armagh reflector. Grubb was in frequent contact with the third Earl of Rosse during the period when he was constructing his 36- and 72-inch reflecting telescopes, and Rosse copied Grubb's system of triangular balanced levers for mounting the primary mirror from the Armagh reflector. Thomas Grubb also made an eye-piece micrometer for the 72-inch telescope, which survives at Birr[101], and instruments were purchased for other Irish observatories[102]. Howard Grubb was similarly involved in Irish scientific life: for example, he was the leader of a Royal Dublin Society Committee which attempted to make the public clocks in Dublin show the correct time[103].

The Grubbs were perhaps the one Dublin firm of instrument makers whose products caught the public imagination. A popular magazine proclaimed, "Visitors to Dublin are certain of finding whole-hearted joyous hospitality on every hand – and big telescopes on the Rathmines road"[104].

Astronomical photography

In 1885 Howard Grubb completed a 20-inch reflector for Isaac Roberts, an English amateur astronomer who wanted to photograph nebulae[105]. The photographic plates available were slow by modern standards, so Roberts required exposures of about four hours for most of the nebulae in which he was interested. The instrument which Grubb supplied had neither a sufficiently regular drive nor an adequately rigid structure and mounting for exposures of this length, and the telescope had to be returned to Dublin for extensive improvements. Although this initial incursion into photographic astronomy caused Grubb many difficulties and must have brought a substantial financial loss, in the longer term it brought a more important contract. Around 1890 he built seven of the 13-inch photographic reflectors for the international *Carte du Ciel* project[106]. He also built a number of other early photographic telescopes, including the 28-inch at Greenwich.

In 1882 the Scottish astronomer, David Gill, then director of the Royal Observatory at the Cape of Good Hope, was working with a commercial photographer to produce images of a comet[107]. He noticed that his plates contained good images of a large number of stars, and realised that it might be possible to use photography to chart the heavens. With the Dutchman, Julius Kapteyn, Gill then organised the Cape Photographic Durchmusterung, which was carried out between 1885 and 1890. This was a photographic survey of over 450,000 stars down to magnitude 9.5. The plates were measured and the resulting catalogue published[108].

The evident success of Gill's work, and the important photographic researches of the brothers Paul and Prosper Henry of Paris into the structure of the Pleiades, led to an Astrographic Congress being held in Paris in 1887[109]. The Congress initiated the *Carte du Ciel*. The project aimed to produce a photographic map of the heavens to fourteenth magnitude, and a catalogue of stars to eleventh magnitude. "The photographic refractor developed by the Henry brothers was adopted as the standard instrument. The specifications called for a 34 cm objective with a focal length of 3.43 m (f/10) so that one millimetre of the photographic plate represented one arc minute on the sky. The 16cm square plates covered a field 2° square. The optics were to be corrected with reference to the Fraunhofer G line. Guiding was accomplished with a 24.5 cm refractor of the same focal length as the photographic telescope and the steel tubes of both instruments were bound together to minimize flexure"[110].

6.11. Sir Howard Grubb, 1894. Trustees of the National Museums of Scotland.

> The almost universal adoption of the photographic method for astronomical observations has introduced a totally new set of conditions into the necessary instrumental equipment of observatories, where satisfactory results are desired. This has in many cases necessitated the remodelling of the designs of instruments that have been, up to the present time, "standard". It has rarely been found possible to remodel existing instruments to fulfil the new conditions, which are much more exacting than the old, and the best results can only be obtained with those instruments which have been constructed from the very beginning with a view to the special work required of them[111].

Howard Grubb was not a lens designer of the ability of his contemporary, H. Dennis Taylor of Thomas Cooke & Sons, and the lenses for the astrographic telescopes were constructed by a process of trial and error, the trials being carried out in Oxford[112]. (As already mentioned, the Vienna achromat had been designed by G.G. Stokes.) The production of the *Carte du Ciel* astrographs required day-and-night working at Rathmines Road for 18 months[113] before they were successfully completed.

The Grubbs were also interested in photography for itself. The detailed history of photography in Ireland has yet to be written, although a number of brief chronologies have been outlined[114]. Thomas Grubb was one of the members of the first committee of the Dublin Photographic Society, formed in 1854 with Sir Joseph Coghill as President. The following year he took out a patent for "an improved photographic lens"[115]. The Dublin Photographic Society became the Photographic Society of Ireland in 1848, and in due course Howard Grubb became a member, and was its President in 1888 and 1889[116]. When the Photographic Convention of the United Kingdom met in Dublin in 1894 for its ninth annual meeting, Sir Howard Grubb was appointed President (*Fig.* 6.11) and, in his address, outlined the aspects of photography which held a particular personal fascination: stereoscropy and astronomical photography[117].

Conclusion

The apotheosis of the independent firm of Howard Grubb coincided with the structural changes in the scientific instrument industry which were described at the end of Chapter Four. There is a feeling of the transitory nature of all achievement in the demise of telescope making in Ireland. Yet Grubb had probably lost control of his firm some time earlier. The facts are not as clear as they might be. It is certain that Grubb was involved in the design of gunsights and submarine periscopes, and that the manufacture of these and other military and naval instruments expanded during the First World War[118]. Indeed, the first submarine periscope in the world was made in Dublin – and by high irony it was placed in one of the submarines designed by John Phillip Holland, the IRA supporter who had fled to the United States, where he designed vessels which he hoped would be used to sink British shipping[119]. Sir Howard Grubb first took out patents concerning the sighting apparatus for ordnance in 1900, and produced many more in the years leading up to and during the First World War[120]. Although the main thrust of the company's activities must have swung over to war-work, it is difficult to estimate how large this was; there seems, in any case, to have been some collaboration with other optical manufacturers, such as Vickers (who were effectively owners of Cookes of York) and R. & J. Beck Ltd in the design of gunsights[121]. The British government were aware of the vulnerability of these supplies: most could probably have been made only by one other company in the country – Barr & Stroud of Glasgow. After 1916 Grubb's works were removed to St Albans, and the transfer was completed by 1918. There is a sense of exile in the move; and Howard Grubb evidently understood it[122]. In retirement he returned to Ireland, and died at Monkstown in 1931.

From an Irish standpoint, one of the Grubbs' greatest triumphs was posthumous. In 1919 Arthur Eddington took to Brazil the astrographic telescope which Greenwich Observatory had bought from Howard Grubb in 1890[123]. With it, and with a coelostat and long-focus lens owned by the Royal Irish Academy and previously used in C.J. Joly's eclipse expedition to Spain in 1900, Eddington performed the first experimental verification of Einstein's Theory of Relativity, by detecting the shift in the position of a star as its light passed close to the Sun at a total eclipse, a shift produced by the gravitational deflection of the light by the Sun. Relativity had been demonstrated, the certainty of science was fading . . . is this not a suitable end to a history?

References and Notes for Chapter Six

1. Patrick Moore, *The astronomy of Birr Castle*, London, 1971. See also J.A. Bennett and Michael Hoskin, The Rosse Papers and Instruments, *Journal of the History of Astronomy*, **12**, 1981, 216-229.
2. H.C. King, *The history of the telescope*, London, 1955, 206-207.
3. J.L.E. Dreyer, *An historical account of Armagh Observatory*, Armagh, 1883, 8.
4. T.R. Robinson, The lay of the speculum [1845] Trinity College Dublin, MS 2225.
5. Moore, *op. cit.*, 2 (ref. 1).
6. Lord Oxmantown, Account of a series of experiments on the construction of large reflecting telescopes, *Edinburgh Journal of Science*, new series, **2**, 1830, 136-144; reprinted in *The scientific papers of William Parsons, third Earl of Rosse, 1800-1867*, [London?],

1926, 8-13 [Henceforth referred to as *Scientific Papers*].

7. *Scientific Papers*, *op. cit.*, 207-221 (ref. 6).
8. King, *op. cit.*, 124 (ref. 2). See also J.A. Bennett, The giant reflector 1770-1870, in E.G. Forbes (ed.), *Human implications of scientific advance*, Edinburgh, 1978, 553-558.
9. J.A. Bennett, "On the power of penetrating into space": the telescopes of William Herschel, *Journal of the History of Astronomy*, **7**, 1976, 75-108.
10. King, *op. cit.*, 129-133 (ref. 2).
11. Lord Oxmantown, Account of a new reflecting telescope, *Edinburgh Journal of Science*, **9**, 1828, 25-30, reprinted in *Scientific Papers*, *op. cit.*, 1-4 (ref. 6).
12. Robinson, *op. cit.*, (ref. 4).
13. King, *op. cit.*, 80, 124 (ref. 2).
14. Lord Oxmantown, *op. cit.* (ref. 6); and T.R. Robinson, Account of the three feet telescope, *Proceedings of the Royal Irish Academy*, **2**, 1840, 2-12; *Scientific Papers*, *op. cit.*, 14-19 (ref. 6).
15. King, *op. cit.*, 189-191 (ref. 2).
16. Lord Oxmantown, An account of experiments on the reflecting telescope, *Philosophical Transactions of the Royal Society*, **130**, 1840, 508-528; *Scientific Papers*, *op. cit.*, 80-104 (ref. 6).
17. William Nicol, On a method of so far increasing the divergency of the two rays in calcareous spar that only one image may be seen at a time, *Edinburgh New Philosophical Journal*, **6**, 1829, 83-84.
18. See for instance, G. L'E. Turner, The rise and fall of the jewel microscope, 1824-1837, *Microscopy*, **31**, 1968, 85-94, and A. Frank and R.H. Nuttall, Makers of jewel lenses in Scotland in the early nineteenth century, *Annals of Science*, **30**, 1973, 407-416.
19. T.R. Robinson, *op. cit.*, (ref. 14).
20. Lord Oxmantown, Account of an apparatus for grinding and polishing the specula of reflecting telescopes, *Edinburgh Journal of Science*, **9**, 1828, 213-217; *Scientific Papers*, *op. cit.*, 5-7 (ref. 6).
21. Lord Oxmantown, *op. cit.*, (ref. 16).
22. *Ibid.*
23. Robinson, *op. cit.*, (ref. 4).
24. For the life of William Parsons, third Earl of Rosse, see A.M. C[larke], *Dictionary of national biography*, **43**, 1895, 425-427.
25. Moore, *op. cit.*, 27 (ref. 1). C.B.W. Smith, *The Great Hunger*, London, 1962, remains a readable if traumatic account of the Famine and its effect on the population.
26. Agnes M. Clarke, *A popular history of astronomy during the nineteenth century*, 4th edn., London, 1908, 118.
27. *Ibid.*, 119.
28. Moore, *op. cit.*, 78 (ref. 1).
29. Inventory no. 1914-370.
30. Robinson, *op. cit.*, (ref. 4).
31. For the life of Burton, see *Monthly Notices of the Royal Astronomical Society*, **43**, 1833, 159-160, and A.P. Fitzgerald, Charles Edward Burton 1846-1882, *Irish Astronomical Journal*, **5**, 1959, 167-173. We are indebted to Professor G.L. Herries Davies for the latter reference.
32. *Monthly Notices of the Royal Astronomical Society*, **43**, 1883, 159.
33. W.E[rck?], The late C.E. Burton, *Astronomical Register*, **20**, 1882, 173-174.
34. King, *op. cit.*, 262-274 (ref. 2).
35. S.M.P. McKenna, Astronomy in Ireland since 1780, *Vistas in Astronomy*, **9**, 1967, 294; C.J. Butler, Early photoelectric photometry in Dublin and Daramona, *Irish Astronomical Journal*, **17**, 1986, 373-277; Ian Elliott, The Monck plaque, *ibid.*, **18**, 1987, 122-124.
36. Moore, *op. cit.*, 51-60 (ref. 1).
37. W.E. Wilson, The effective temperature of the sun, *Monthly Notices of the Royal Astronomical Society*, **62**, 1902, Appendix, 64-72.
38. King, *op. cit.*, 256 (ref. 2).
39. C.E. Burton, Refractors and reflectors compared, *Astronomical Register*, **10**, 1872, 289-290.
40. George Johnstone Stoney, On a new form of spectroscope, *Proceedings of the Royal Irish Academy*, 2nd series, **1**, 1874, 208-212.
41. C.E. Burton and Howard Grubb, On a new form of ghost micrometer for use with astronomical telescopes, *Scientific Proceedings of the Royal Dublin Society*, **3**, 1881, 1-11.
42. C.E. Burton, On a spectroscope of the binocular form for the observation of faint spectra, *Proceedings of the Royal Irish Academy*, **2**, 1877, 42-45. The instrument is now in the Science Museum, London, inventory no. 1900-141.
43. C.E. Burton, Note on the spectrum, polarization, and form of the zodiacal light; as observed in the years 1874 and 1875, *Proceedings of the Royal Irish Academy*, **2**, 1877, 218-223.
44. *Burke's Irish family records*, London, 1976, 515-527.
45. Geoffrey Watkins Grubb, *The Grubbs of Tipperary*, Cork, 1972, 85-86.
46. For the life and work of Thomas Grubb, see A.M. C[larke], *Dictionary of national biography*, **23**, 1890, 301; H.C. King, in C.C. Gillispie (ed.), *Dictionary of scientific biography*, **5**, 1972, 561-562; and King, *op. cit.*, 265-267 (ref. 2); and the obituaries in *Observatory*, **2**, 1878, 203; *Nature*, **18**, 1878, 570; *Athenaeum*, no. 2658, 1878, 435; *Proceedings of the Royal Irish Academy*, 2nd series, **3**, 1880, 70.
47. W.G. Fitzgerald, Sir Howard Grubb, FRS,

FRAS, etc. etc., *Strand Magazine*, **7**, 1896, 369-381.

48. For the life and work of Howard Grubb, see H.C. King, in C.C. Gillispie (ed.), *Dictionary of scientific biography*, **5**, 1972, 561; H.C. King, *op. cit.*, 292, 299-301, 306, 350-351, 431 (ref. 2); and the obituaries in *Proceedings of the Royal Society*, **135A**, 1932, iv-ix; *Monthly Notices of the Royal Astronomical Society*, **92**, 1932, 253-255; *Nature*, **128**, 1931, 609-610.
49. Fitzgerald, *op. cit.*, 369 (ref. 47).
50. Sir Howard Grubb, Parsons & Co., *Astronomical and other instruments*, Newcastle-upon-Tyne, 1926.
51. Albert van Helden, Telescope building, 1850-1900, in Owen Gingerich (ed.), *Astrophysics and twentieth-century astronomy to 1950*, The General History of Astronomy, vol. 4A, Cambridge, 1984, 40-58.
52. Recently the topic of two books: Helen Wright, *James Lick's monument: the saga of Captain Richard Floyd and the building of the Lick Observatory*, Cambridge, 1987, and D.E. Osterbrock, J.R. Gustafson and W.J. Shiloh Unruh, *Eye on the Sky*, Berkeley, 1988.
53. H.C. Vogel, quoted in Howard Grubb, *An illustrated catalogue of astronomical instruments, observatories, etc. manufactured by Sir Howard Grubb, F.R.S., F.R.A.S., Etc., Astronomical Instruments Works, Rathmines, Dublin*, Dublin, 1888, 4.
54. Charles Pritchard, The University Observatory at Oxford, *Monthly Notices of the Royal Astronomical Society*, **36**, 1875-1876, 1-5.
55. Obituary of Howard Grubb, *Monthly Notices* . . . (ref. 48).
56. King, *op. cit.*, 312 (ref. 2).
57. Howard Grubb, *Astronomical instruments, observatories, etc., manufactured by Sir Howard Grubb*, Dublin, 1908, 19-20.
58. Howard Grubb, *op. cit.*, 25 (ref. 53).
59. Anon., Astronomical telescopes at the Manchester Exhibition, *Engineering*, **44**, 1887, 626, 630-632, 667-668.
60. Howard Grubb, *op. cit.*, 6 (ref. 57).
61. *Ibid.*, 2.
62. *Ibid.*, i.
63. Thomas Grubb, On a new table microscope, *Journal of the Royal Dublin Society*, 3, 1860, 85-88. This article was reprinted as Grubb's sector microscope, *Journal of the Royal Microscopical Society*, 3, 1880, 1055-1059, and in the *English Mechanic*, **31**, 1880. Thomas Grubb took out a provisional patent, no. 1477, 5 July 1854, for "Improvements in microscopes", and the main features were described by him in On an improvement in the illumination of objects in the compound microscope, *Proceedings of the Royal Irish Academy*, **5**, 1853, 296-297. There is a Grubb microscope according to this design in the Museum of the History of Science, Oxford, inventory no. 67-65.
64. Howard Grubb, On Babbage's system of mechanical notation as applied to automatic machinery, *Scientific Proceedings of the Royal Dublin Society*, **1**, 1878, 111-116.
65. Howard Grubb, Notes on some points in the construction of turret clocks, *ibid.*, **4**, 1885, 447-448.
66. Howard Grubb, Revolving machinery for the domes of astronomical observatories, *ibid.*, **7**, 1892, 484-491.
67. Sir Howard Grubb, Parsons & Co., *op. cit.*, 44 (ref. 50).
68. Howard Grubb, *op. cit.*, 22 (ref. 57).
69. Award of the Boyle Medal to Sir Howard Grubb, F.R.S., 1912, *Scientific Proceedings of the Royal Dublin Society*, **13**, 1912, 288-292.
70. William Lassell had built his great instrument in 1859-60: see J.A. Bennett, A surviving flat from William Lassells's four-foot equatorial Newtonian telescope, *Journal of the History of Astronomy*, **12**, 1981, 195-197; also discussed in Val Helden, *op. cit.*, 54-55 (ref. 51).
71. Howard Grubb, *Report of the forty-sixth meeting of the British Association, held at Glasgow, in September 1876*, sectional reports, 36-37. Italics in the original.
72. Howard Grubb, On the equatorial telescope, and on the new observatory of the Queen's College, Cork, *Scientific Proceedings of the Royal Dublin Society*, 2, 1880, 347-369, especially 361-364.
73. King, *op. cit.*, 431 (ref. 2).
74. Howard Grubb, On a new form of equatorial telescope, *Scientific Transactions of the Royal Dublin Society*, **3**, 1884, 61-68.
75. Samuel Hunter, On the various forms of apparatus used for polishing specula for reflecting telescopes, *ibid.*, **1**, 1878, 97-109. The article "Speculum" (written by T.R. Robinson) in J.P. Nichol, *A cyclopaedia of the physical sciences*, 2nd edn., London, 1860, 779-790 describes and illustrates Grubb's earlier machine.
76. Willem Hackmann, The nineteenth-century trade in natural philosophy instruments in Britain, in P.R. de Clercq (ed.), *Nineteenth century scientific instruments and their makers*, Leiden, 1985, 53-91.
77. Obituary of R.W. Munro, *Quarterly Journal of the Royal Meteorological Society*, **39**, 1913, 72.
78. Sir Howard Grubb, Parsons & Co., *op. cit.*, 35 (ref. 50).
79. The information in this section is from two

sources: Sir Howard Grubb's Works, Dublin, *Engineering*, **46**, 1888, 571-573; Fitzgerald, *op. cit.*, (ref. 47).

80. [Charles Piazzi Smyth?], The Royal Observatory, Edinburgh, *Monthly Notices of the Royal Astronomical Society*, **33**, 1872-73, 221.
81. Grubb, *op. cit.*, 8 (ref. 57).
82. King, *op. cit.*, 306 (ref. 2). See also Anon., The Vienna Equatorial, *Engineering*, **29**, 1880, 7, 114-116, 199-202, 309-311, 409, 467-469; *ibid.*, **30**, 1880, 314-315, 424-428.
83. *Memoir and correspondence of the late Sir George Gabriel Stokes*, 2 vols, Cambridge, **1**, 1906, 357-358.
84. *Ibid.*, **1**, 205-207.
85. *Monthly Notices*, *op. cit.*, (ref. 48).
86. University College Dublin, Archives RCSI/263, ff.1-11.
87. John Burnett, The origins of the electrocardiograph as a clinical instrument, in W.F. Bynum, C.J. Lawrence and V.F. Nutton (eds.), *The emergence of modern cardiology*, London, 1985, 53-76; M.J.C. Cattermole and A.F. Wolfe, *Horace Darwin's shop: the Cambridge Scientific Instrument Company 1878-1968*, Bristol, 1987.
88. T.N. Clarke, A.D. Morrison-Low and A.D.C. Simpson, *Brass & glass: scientific instrument making workshops in Scotland as illustrated by instruments from the Frank Collection at the Royal Museum of Scotland*, Edinburgh, 1989.
89. Obituary of Archibald Barr, *Nature*, **128**, 1931, 294-295. See also M. Moss and I. Russell, *Range and vision: the first hundred years of Barr & Stroud*, Edinburgh, 1988.
90. He would have received this award earlier, but for a rule that honorary officers of the Royal Dublin Society were not eligible for it; Award . . ., *op. cit.* (ref. 69).
91. C. Mollan, *Nostri plena laboris* – an author index to the RDS scientific journals – 1800-1985, Dublin 1987; H.F. Berry, *A history of the Royal Dublin Society*, London, 1915; J. Meenan and D. Clarke (eds.), *The Royal Dublin Society 1731-1981*, Dublin 1981.
92. *Burke's* . . ., *op. cit.*, 518 (ref. 44).
93. Obituary of Thomas Romney Robinson, *Monthly Notices of the Royal Astronomical Society*, **42**, 1882-1883, 181-183.
94. Mercator [i.e. R.L. Edgeworth], *Monthly Magazine*, **12**, 1801, 191-192.
95. Humphrey Lloyd, On the declination instrument employed in the magnetical observatory of Dublin; an Account of the magnetical observatory of Dublin, *Proceedings of the Royal Irish Academy*, **1**, 1841, 163-166, 330-334; Anita McConnell, Nineteenth-century geomagnetic instruments and their makers, in de Clercq, *op. cit.*, 29-52 (ref. 76).
96. Science Museum, inventory no. 1876-796. See Anita McConnell, *Geophysics and geomagnetism: catalogue of the Science Museum collection*, London, 1986, 33.
97. Ann Savours and Anita McConnell, The history of the Rossbank Observatory, Tasmania, *Annals of Science*, **39**, 1982, 527-564.
98. Trinity College Dublin, MUN/P/4/218/26; 220/25, 25a; 231/32, 32a; 235/21, 21a.
99. Trinity College Dublin, MUN/P/4/238/24, 24a; 261/36, 36a, 36b. A magnetometer by Thomas Jones, and dip circles by Robinson & Barrow and by Gambey, remain in the collection of the Physics Department of Trinity College, Dublin. We are grateful to Dr Charles Mollan for this information.
100. Trinity College Dublin, MUN/P/4/219/36, 36a, 52.
101. Bennett and Hoskin, *op. cit.*, 218 (ref. 1).
102. See, for instance, Patrick A. Wayman, The South Telescope at Dunsink Observatory, *Irish Astronomical Journal*, **8**, 1968, 274-276, Patrick A. Wayman, Rev. John Brinkley and Sir James South, *ibid.*, **17**, 1986, 378-382; Patrick A. Wayman, *Dunsink Observatory 1785-1985, a bicentennial history*, Dublin, 1987; Mackenna, *op. cit.*, (ref. 35); we are grateful to Professor Wayman for these references.
103. Howard Grubb, On the correction of errors in the distribution of time signals, *Scientific Proceedings of the Royal Dublin Society*, **9**, 1899-1902, 37-45; Award . . . (ref. 69). See also Wayman, *Dunsink . . .*, *op. cit.*, 127-154 (ref. 102).
104. Sir Howard Grubb and his work, *Illustrograph*, no. 34, 1896, 13.
105. Isaac Roberts, *A selection of photographs of stars, star clusters and nebulae*, London 1893-99, 21.
106. Sir Howard Grubb, Parsons & Co., *op. cit.*, 45 (ref. 50). The telescopes went to Greenwich, Oxford, Melbourne, Cork, Perth (Western Australia), Cape Town, and Tacubaya (Mexico).
107. George Forbes, *David Gill: man and astronomer*, London, 1916, 134-136; John Lankford, The impact of photography on astronomy, in Gingerich, *op. cit.*, 16-39 (ref. 51), especially 25-33.
108. The results were published as the Cape Photographic Durchmusterung in *Annals of the Cape Observatory*, **3-5**, 1896-1900.
109. H.H. Turner, *The great star map: being a brief general account of the international project known as the Astrographic Chart*, London, 1912; P. Couderick, Historique de la Commission de la Carte du Ciel, *Transactions of the*

International Astronomical Union 1970, **14B**, 1971, 170-176.

110. Lankford, *op. cit.*, 29 (ref. 107).
111. Howard Grubb, *op. cit.*, 3 (ref. 53).
112. King, *op. cit.*, 300 (ref. 2).
113. Fitzgerald, *op. cit.*, 373 (ref. 47).
114. Edward Chandler, *Photography in Dublin during the Victorian Era*, Dublin, n.d. [1982]; Edward Chandler, Photography, in B. de Breffney, *Ireland: a cultural encyclopaedia*, London, 1983, 192-193.
115. Patent no. 2574, 8 October 1857.
116. Oscar S. Merne, *The story of the Photographic Society of Ireland 1854-1954*, Dublin, 1954, 4.
117. [H. Grubb], The President's Address, *British Journal of Photography*, **41**, 1894, 438-439.
118. Howard Grubb, A new collimating telescope gun-sight for large and small ordnance, *Scientific Transactions of the Royal Dublin Society*, 7, 1901, 321-330; Anon., Grubb's collimating telescope gun-sight, *Engineering*, **71**, 1901, 619; Anon., Gun-sights for large and small ordnance, *Nature*, **65**, 1902, 226-228.
119. J. de Courcy Ireland, *Ireland and the Irish in maritime history*, Dun Laoghaire, 1986, 262-267.
120. Patents no. 12,108, 4 July 1900; no. 20,813, 17 November 1900; no. 22,127, 5 December 1900; no. 5218, 1 March 1902; no. 5806, 19 March 1902; no. 8735, 27 April 1901; no. 10,373, 18 May 1901; no. 21,856, 10 October 1903; no. 25,682, 24 November 1903; no. 13,438, 24 June 1908; no. 4005, 16 February 1911; no. 25,476, 6 November 1912; no. 1511, 18 January 1913; no. 22,319, 10 November 1914; no. 15,662, 5 November 1915; no. 12,445, 30 August 1915; no. 103,178, 20 January 1916; no. 124,497, 7 February 1916; no. 157,478, 10 March 1919; no. 161,214, 6 November 1919; no. 178,474, 13 October 1920. Most of these related to improvements in the optics of gunnery. Not all of them were taken out solely in Sir Howard Grubb's name.
121. For instance the periscopes outlined in patents no. 1511, 18 January 1813 and no. 22,319, 10 November 1914; and various optical systems in 1915 and 1916.
122. For further evidence of Howard Grubb's patriotic feelings towards Ireland, see The Late Sir Howard Grubb, *Engineering*, **132**, 1931, 403.
123. Described in Arthur Eddington, Forty years of astronomy, in J. Needham and W. Pagel (eds.), *Background to modern science*, Cambridge, 1940, 140-142.

APPENDIX

The leading instruments built by Thomas and Howard Grubb, 1834-1926

The sources used for this survey, which is by no means complete, were the *Monthly Notices of the Royal Astronomical Society*; *Journal for the History of Astronomy*; *Engineering*; C. André et G. Rayet, *L'Astronomie pratique et les observations en Europe et en Amerique*, 5 vols, Paris, 1874; P. Stroobant *et al.*, *Les observatoires astronomiques et les astronomes*, Brussels, 1907; A.M. Clarke, *A popular history of astronomy in the nineteenth century*, 4th edn., London, 1907; André Danjon et André Couder, *Lunettes et telescopes*, Paris, 1935; George Z. Dimitroff and James G. Baker, *Telescopes and accessories*, London, 1946; M.C. Donnelly, *A short history of observatories*, Eugene, Oregon, 1973; P. Muller, *Sternwarten: Architektur und Geschischte der Astronomischen Observatorien*, Frankfurt, 1975. We are grateful to D.J. Bryden for the last two references. Dates over which there is a significant uncertainty are given in square brackets.

1.	1834	13.3″ refractor	Markree, Co. Sligo	For E.J. Cooper. Object glass by Cauchoix of Paris, permanent mounting by Thomas Grubb. Sold to the Jesuit Seminary at Aberdeen, Hong Kong, c.1928; bombed 1941; sent to Manila Observatory, the Philippines, c.1947.
2.	1836	15″ reflector	Armagh	Newtonian/Cassegrain equatorial mounting.
3.	1838	6.7″ refractor	Royal Observatory, Greenwich	Mounting only: object glass by Cauchoix of Paris. Known as the "Sheepshanks Equatorial". National Maritime Museum, inventory no. 00/R.9.
4.	[c.1840-1856]	6″ refractor	U.S. Military Academy, West Point	Equatorial mounting only: object glass by Lerebours of Paris.

5.	1853	22″ reflector	Glasgow University	22″ mirror made by Ramage of Aberdeen, c.1843. Reworked by Grubb 1855, paid for by the Marquis of Breadalbane. Subsequently known as the "Breadalbane telescope", New 20″ mirror by Grubb, 1894.
6.	1868	12″ refractor	Dunsink	Object glass by Cauchoix, c.1830, for Sir James South, who presented it to the University of Dublin in 1863; dome by Grubb, 1894.
7.	1869	48″ reflector	Melbourne	Cassegrain, metal mirror. Mounting moved to Mt Stromolo in 1954.
8.	1870	15″ refractor and 18″ reflector	Tulse Hill, London	Purchased by the Royal Society and lent to William Huggins. Cassegrain reflector, metal mirror. Given to the Solar Physics Laboratory, Cambridge, in 1910.
9.	1871	12″ reflector	Daramona, Co. Westmeath	Purchased by W.E. Wilson.
10.	1872	24″ reflector	Royal Observatory, Edinburgh	Designed by Grubb and C.P. Smyth; advice on the mirror from the Rev. Henry Cooper Key; installed 1873 but it and dome unusable. Returned to Grubb, re-installed in 1886. Moved from the Calton Hill site to Blackford Hill, Edinburgh, 1896.
11.	1873	15″ refractor	Dun Echt, Aberdeenshire	Given to the Royal Observatory, Edinburgh, by the Earl of Crawford in 1894. Reconditioned by Sir Howard Grubb, Parsons & Co., Newcastle, 1929.
12.	1874	13″ reflector	Dun Echt, Aberdeenshire	Newtonian/Cassegrain reflector, metal mirror, also silver glass mirror. Used by Lord Lindsay on Transit of Venus expedition to Mauritius. Transferred to the former Corporation of Edinburgh in 1896; parts now in the Royal Museum of Scotland.
13.	1875	12.25″ refractor	University Observatory, Oxford	
14.	1878	8″ refractor	University College, Cork	Also multiprism spectroscope c.1880.
15.	c.1878	Heliometer	Royal Observatory, Cape of Good Hope	Bought by Lord Lindsay, for observations of Juno and Mars, and sold by him to the Cape.
16.	1881	27″ refractor	University Observatory, Vienna	Also three observatory domes by Grubb.
17.	1881	8″ refractor	Castlemaine, Victoria	Dr W. Bone's private observatory.
18.	1881	24″ reflector	Daramona, Co. Westmeath	For W.E. Wilson. Remounted 1891. Moved to Mill Hill Observatory, London, now at the National Museums on Merseyside.
19.	[pre-1882]	8″ refractor	Royal Observatory, Potsdam	
20.	1882	Compound prism spectroscope	Dun Echt, Aberdeenshire	
21.	1882	8″ refractor	Yale University Observatory, New Haven, Connecticut	Funds anonymously donated by Edward Mordecai Reed, so known as the "Reed Telescope".

22.	1885	20″ reflector	Maghull, near Liverpool	With 7″ refractor on the same mounting. Bought by Isaac Roberts. Moved to Crowborough, Sussex, in 1890. Now in the Science Museum, London, inventory no. 1936-231.
23.	1885	10″ refractor	Armagh	Papier-maché dome also by Grubb.
24.	1886	9″ photographic refractor	Royal Observatory, Cape of Good Hope	Objective presented by James Nasmyth, with 6" Dallmeyer rapid rectilinear lens and 5" visual guider mounted by Grubb to a design by H.M. Astronomer.
25.	1888	Sidereal clock	Rousdon Observatory, Devon	Private observatory, built 1884, by Sir Cuttibent E. Peek, Bt.
26.	1890	13″ refractor	Royal Observatory, Greenwich	With 11″ visual guider. For the *Carte du Ciel* project. Now Royal Greenwich Observatory, Dome D.
27.	1890	13″ refractor	Melbourne	With 11″ visual guider. For the *Carte du Ciel* project.
28.	1890	15.4″ (380mm) equatorial	National Mexican Observatory, Tacubaya	
29.	c.1890	13″ refractor	National Mexican Observatory, Tacubaya	With 11″ visual guider. For the *Carte du Ciel* project.
30.	c.1890	13″ refractor	University Observatory, Oxford	With 11″ visual guider. For the *Carte du Ciel* project.
31.	c.1890	13″ refractor	Cork	With 11″ visual guider. For the *Carte du Ciel* project.
32.	c.1890	13″ refractor	Perth, Western Australia	With 11″ visual guider. For the *Carte du Ciel* project.
33.	c.1890	13″ refractor	Royal Observatory, Cape of Good Hope	With 11″ visual guider. For the *Carte du Ciel* project.
34.	c.1890	9″ photographic refractor	Hampton, Middlesex	For Sir Henry Thompson's observatory.
35.	c.1890	20-foot dome	Royal Observatory, Göttingen	
36.	c.1890	8″ refractor	Caracas, Venezuela	15-foot observatory dome.
37.	1891	9″ photoheliograph	Royal Observatory, Greenwich	
38.	1893	28″ refractor	Royal Observatory, Greenwich	Visual and photographic instrument; object glass only. The mounting had been made by Ransome & Simm in 1859. Royal Greenwich Observatory, Dome F.
39.	1893	15″ refractor	Maidenhead, Berkshire	Photographic doublet. Also dome and lifting floor. By 1926 at Hepple Woodside, Northumberland.
40.	1893	15″ refractor	University of Mississippi Observatory	Visual; mounted with a 9″ photographic refractor.
41.	1894	15″ refractor	Stonyhurst College, Lancashire	*In memoriam* Fr. Perry, S.J., who died in British Guiana on the Total Solar Eclipse Expedition, 22 December 1889.

42.	1895	36″ reflector	Lick Observatory, University of California	The "Crossley" reflector, made by A.A. Common in 1879, refigured by Grubb.
43.	c.1896	12″ refractor	Tunbridge Wells, Kent	For Frank McClean, 10 degree objective prism. Subsequently gifted to Sir Norman Lockyer's observatory at Silcombe Regis, Exeter, and mounted with a 9″ refractor.
44.	1896 1897	30″ reflector 26″ photographic refractor	Royal Observatory, Greenwich	Given by Sir Henry Thompson; known as the "Thompson Equatorial". Reflector mirror by Common; instruments mounted together while at Greenwich. Royal Greenwich Observatory, Dome A.
45.	1897	10″ refractor	Coats Observatory, Paisley	Instrument still *in situ*. Bifilar micrometer now in Royal Museum of Scotland, inventory no. T1984.22.
46.	[1898-1907]	20″ reflector	Maharaja Takhtasingji Observatory, Poona.	Mounting only: mirror by A.A. Common, Ealing, London. Cassegrain equatorial, with spectrographs, polarising prism and positional micrometer by Grubb.
47.	[pre-1899]	12″ refractor	Dresden	Dr Engelhardt's observatory. Instruments given by him to the Imperial University of Kazan, Russia, in 1899.
48.	1899	12.5″ coudé	Solar Physics Observatory, Cambridge	Triple photographic/visual; objectives by Cooke & Sons, York; mirror by Common. Design and mounting by Grubb. Known as the "Sheepshanks Telescope".
49.	1900	16″ refractor	Grand-Ducal Observatory, Heidelberg	Photographic doublet. Lenses by Schott of Jena, finished by Brashear of Pittsburg; mounted by Grubb. The gift of Miss Bruce. Mounted with a visual 10″ refractor by Pauly.
50.	[c.1900]	8″ equatorial	Natal Observatory, Durban	
51.	[c.1900]	8″ equatorial	Private observatory, Louvain, Brussels	
52.	[c.1900]	16″ equatorial	Astronomical Observatory, Madrid	
53.	[c.1900]	Equatorial astrograph	University Observatory, Oxford	
54.	[c.1900]	Spectrograph and sideostat	Kodaikanal Observatory, Palani Hills, Madras, India	
55.	[c.1900]	8″ equatorial	The Peninsula, Windsor, New South Wales	Private observatory of John Tebbutt. [Founded 1864.]
56.	1901	24″ photographic refractor	Royal Observatory, Cape of Good Hope	Given by Frank McClean; known as the "Victoria" telescope. Mounted with an 18″ visual refractor, and with spectrographic apparatus.
57.	1902	24″ double refractor	Radcliffe Observatory, Oxford	Photographic, with 18″ visual lens. Also 32-foot dome and rising floor. Moved to Mill Hill in 1931.
58.	1903	15″ refractor	Nizamiah Observatory, Hyderabad	

59.	c.1905	12-prism spectroscope	Cartuja, Granada, Spain	
60.	c.1905	24″ refractor	National Astronomical Observatory, Santiago, Chile	Also 45-foot dome and rising floor.
61.	c.1905	20″ refractor	Bombay	
62.	1924	7m solar spectrograph	Pulkovo Observatory, Simeis, Russia.	
63.	1925	40″ reflector	Pulkovo Observatory, Simeis, Russia.	
64.	1925	26″ refractor	Union Observatory, Johannesburg, South Africa.	

CHAPTER SEVEN

DUBLIN INSTRUMENT MAKERS

A.D. Morrison-Low

One of the tragedies of Irish historiography is the dearth of source material: an enormous quantity of archives was destroyed in the Public Record Office when it came under fire in May 1922. This has meant that we have had to rely on sources other than manuscripts. This listing of Dublin instrument makers has been compiled from Dublin trade directories, together with information gleaned from newspaper advertisements and surviving instruments. The first trade directory for Dublin was that published in 1752 by Peter Wilson, which was produced thereafter annually, except for 1754, and usually bound with John Watson's *Almanack* which had appeared annually since 1729[1]. Wilson's *Directory* survived until 1837; between 1834 and 1848 Messrs Pettigrew and Oulton produced the *Dublin almanac and general register of Ireland*, and the General Post Office issued an annual *Directory* between 1832 and 1843. In 1844 Alexander Thom published the first edition of the *Irish almanac and official directory* which is still in production. In addition, information has also been obtained from composite directories covering larger parts of the country: J. Pigot's *Commercial directory of Ireland . . .*, Manchester, 1820; Pigot & Co.'s *City of Dublin and Hibernian provincial directory*, London, 1824; and I. Slater's *National commercial directory of Ireland*, London, 1836.

Information from trade directories should be treated with caution, as the facts may not be accurate for the date when they appear; for instance, no instrument maker appears in Wilson's *Directory* before 1760, when there is one instrument maker and one weatherglass maker; in 1761 there are two mathematical instrument makers, one optician, one weatherglass maker and one barometer maker. This does not mean that they all started in business in that year; it is more probable that they did not find it worthwhile advertising their presence through the *Directory* in its earliest years. Similarly, old information was sometimes carried over from one year to the next, so that, although we know from family sources that John Alment died in 1787 at the age of 47[2], his name appeared in the *Directory* for the following year.

Another problem is the apparent frequency of change of premises made by some businesses, often within the same street. This may be straightforward typographical error, but the issue was confused by the numbering of properties which began in 1774; vacant sites were not allowed for and this led to occasional renumbering, with the resulting overlapping unresolved until 1834[3].

Given that an appearance in a street directory or in its trades section was a form of advertisement, it is not always evident from the entries whether people were manufacturers of instruments or merely retailers, and other sources have had to be used to confirm this. We have been fortunate in that two lists of Irish instrument makers, compiled in the early part of this century, survive in manuscript form at the Museum of the History of Science, Oxford. One of these lists was compiled by M.D. Westropp, who was for many years Keeper of the Industrial and Art Division at the National Museum of Ireland; the author of the other was Thomas H. Mason (1877-1958) from the family of instrument makers. Unfortunately, neither list gives its sources, but both have proved reasonably reliable where we have been able to check the material. A combined version of these lists was published in the *Museum third Tostal display souvenir catalogue* (St Patrick's College, Maynooth, 1955). We are grateful, too, for the continuing discussion and exchange of information with Project Simon.

As curators, we have turned to another source for information: that of the surviving instruments. These can be found all over Ireland, in various institutions and private collections. Often the three dimensional artefact supplies information which a document cannot, particularly for the early years of instrument use and production. We have found names and addresses of makers who simply are not recorded elsewhere, engraved on pieces; in these cases we have had to use stylistic grounds to date the maker, bearing in mind that instruments from Ireland, similar to those made in Scotland, tended to remain unchanged for longer than those made near the centre of the trade in Southern England.

Dublin instrument makers

ALLEN, Thomas

Died 1767[4]	Mathematical Instrument Maker.

ALMENT, John (c.1740-1787)

1767	Mathematical Instrument Maker, Mary's Abbey[5].
1772-1774	Optician, Mary's Abbey.
1775-1788	Optician, 34 Mary's Abbey.
	Previously foreman to John Margas[6]. Died in 1787, aged 47[7], although his name appears in the directory for 1788.

AUSTIN, John

1818-1820	Mathematical Instrument Maker, 2 Essex Quay.

BARON, Patrick

1794	Optician, 83 Dame Street.

BATE, John

fl. 1750-1763	A brass sundial by John Bate, Dublin, is known, engraved "Capt. Willm. Robt. Adire Ballymenogh lat. 54-55, 1750". Westropp gives his dates as "1739 died 1780"[8]. His address was Nicholas Street in 1756, which implies that Philip was a relative[9]. The International Genealogical Index records a Philip Bate, christened 2 March 1739 at St Nicholas Within, Dublin; parents John and Alice Bate.

BATE, Philip

1761-1762	Mathematical Instrument Maker, Nicholas Street.
	A brass universal equinoctial dial signed "John & Phil Bate Fecit Dublin 1763" was offered for sale by Sotheby's Chester 30 November 1988 Lot 381.

BECKER, T.E. & Co. (sic)

1903-1905	Laboratory Furnishers, 182 Great Brunswick Street.
	Although appearing in the directory as T.E. Becker & Co., this was a branch of the Manchester firm of F.E. Becker & Co.

BOOKER, Francis and John

1761-1763	Glass Grinders, Essex Bridge.
1764-1773	Glass Sellers, Essex Bridge.
	Their trade card illustrates a microscope and telescope[10].

BOOKER, John

1774-1786	Looking-glass Seller, 6 Essex Bridge.
1787-1789	Looking-glass Seller, 4 Jervis Street.
	Also traded in Cork.

BUCKLEY, Joseph

1832-1834	Optician, 14 Lower Sackville Street.
1835-1844	Optician, Mathematical and Philosophical Instrument Maker, 14 Lower Sackville Street.
1845-1859	Optician, 14 Lower Sackville Street.
	A Joseph Buckley married Anna Maria Grieg in 1829[11].

CAHILL, Patrick

1876-1882	Optician, 17½ Wellington Quay.
1883	Optician, 17½ and 20 Wellington Quay.
1884-1886	Optician, 17½ Wellington Quay.
	There seems to be a close connection between this business and that of P.J. Keary, who appears to have shared the same premises.

CAHILL, Patrick, & Co.

1887	Opticians, 17½ Wellington Quay.

CAHILL, Patrick, & Son

1888-1891	Opticians, 17½ Wellington Quay.

CAHILL, Patrick, jnr.

1884-1891	Optician, 17½ Wellington Quay.
1892-1893	Optician, 17½ Wellington Quay and 13 Wellington Quay.
1894	Optician, 17½ Wellington Quay.
1895-1897	Optician and Mathematical Instrument Maker, 17 and 17½ Wellington Quay.
1898	Optician, 13 and 17½ Wellington Quay.
1899-1904	Optician, 13 Wellington Quay.
	The word "junior" is used only in the Directories for 1884 and 1887.

CAHILL, K. Patrick

1905-1907	Optician, 13 Wellington Quay.

CAHILL, Patrick, Ltd

1908-1909	Opticians and Philosophical Instrument Makers, 13 Wellington Quay.
1910-1914	Opticians to his Holiness Pope Pius X, 13 Wellington Quay.
1915	Opticians to his Holiness Pope Pius X, 13 and 15 Wellington Quay.
1916-1918	Optician, 13 and 15 Wellington Quay and 3 Anne Street, South.

CAHILL, Patrick

1919-1922	Optician, 13 and 15 Wellington Quay.

CARTLAND, George

1824	Optician, 8 Rosemary Street.
	Appears only in Pigot and Co.'s *City of Dublin and Hibernian provincial*

directory, London, 1824; does not appear in Wilson's *Dublin directory* for either 1824 or 1825.

CASTELLO, Joseph

1854-1922 Optician, 42 Wellington Quay.
Castello does not appear in the directories for 1893 or 1894. From 1854 to 1859 the directories give his name as Costelloe.

CASTELLO & Son

1891-1894 Opticians, 42 Wellington Quay.

CAVE, Thomas

fl. 1729-1749 Mathematical Instrument Maker and Surveyor.
Andrews[12] notes work by Cave in 1729, and an advertisement of May 1749, by which time he is described as "the late Mr Thomas Cave".

CHANCELLOR, John

1862-1888 Optician and Photographer, 55 Sackville Street Lower.
1889-1898 Optician and Photographer to the Prince of Wales, 55 Sackville Street Lower.

CHANCELLOR & Son

1899-1902 Photographers, Jewellers and Opticians, 55 Sackville Street Lower; Clock Factory, 32 Batchelor's Walk.
1903 Photographers, Jewellers and Opticians, 55 Sackville Steet and 7 Grafton Street.
1904 Watch and Clock Manufacturers and Photographers, Jewellers and Opticians, 55 Sackville Street and 7 Grafton Street.
1905-1922 Watch and Clock Manufacturers and Photographers, 55 Sackville Street, 7 Grafton Street; factory, Adam Court.

CLARKE, Edward

1810-1812 Optician, 18 Lower Sackville Street.
1819-1821 Optician, 10 Lower Sackville Street.
Clarke does not appear in the directories for 1813, 1814 or 1818, but was in partnership with Richard Spear in 1815-1817.

CLARKE, E. & Co.

1823-1831 Optician, 83 Dame Street.
1832 Mathematical and Philosophical Instrument Maker, 83 Dame Street.
The 1822 directory does not contain entries for either Edward Clarke or E. Clarke & Co.

CONNOLLY, George

1835-1844 Optician, 45 Grangegorman Lane.
1845 [no trade], 45 Grangegorman, Upper.

COSTELLO, Joseph: see CASTELLO, Joseph

CRUMPE, William

1876-1882 Optic Warehouse, 2 Abbey Street Lower.
1883-1884 Optician, Jeweller, etc., 2 Abbey Street Lower.
1885-1899 Optician, Jeweller and Cutler, 41 Henry Street.

CURTIS & Co.
1903-1904 Optical and Photographic Contractors to H.M. Government, 65 Dawson Street.

CURTIS BROTHERS
1893-1897 Electrical Engineers, Opticians and Scientific Instrument Makers, 10 Suffolk Street.
The business appears to have been continued as Dobson & Curtis.

DAWSON, John
1883-1884 Optician, 8 Fleet Street.
1885-1886 Optician, 2 Fleet Street.

Del VECCHIO, J.[13]
1810-1817 Printseller and Looking-glass Manufacturer, 26 Westmoreland Street.
1818-1836 Printseller, Looking-glass and Plaster of Paris Manufacturer, 26 Westmoreland Street.
1837-1838 Printseller, Looking-glass and Plaster of Paris Manufacturer, 26 Westmoreland Street and 187 and 188 Great Brunswick Street.
The *Commercial directory* for 1820-1822 lists him as a Barometer Maker, Looking-glass Maker and Cement Manufacturer, 26 Westmoreland Street and 6 D'Olier Street.

Del VECCHIO, James
1833-1835 Looking-glass Manufacturer and Frame Maker, 15 Lower Abbey Street.
1836-1838 Looking-glass Manufacturer and Frame Maker, 68 Dame Street.

DICKSON, Thomas
c.1720 Mathematical Instrument Maker.
"Mentioned 1720"[14].

DIXON & HEMPENSTALL
1904-1906 Opticians, Dealers, etc., 12 Suffolk Street.
1907-1922 Opticians, Mathematical Instrument Makers, etc., 12 Suffolk Street.

DOBSON & CURTIS BROTHERS
1898-1899 Builders' Providers, Painters, Merchants, Decorators and Importers of Paper Hangings, Dealers in Sanitary Goods, Glass Fittings, Wood and Marble Mantelpieces, Electrical Engineers, Optical and Photographic Instrument Makers, 114 Stephens' Green West, Cuffe Lane, Suffolk Street and Birmingham.
1900 Builders' Providers, Painters, Decorators, Electrical Engineers, etc., 114 Stephen's Green West, Cuffe Lane, Suffolk Street, 74 South Mall, Cork, and Bangor, Wales.
1901 Builders' Providers, Painters, Decorators, Electrical Engineers, etc., 114 Stephen's Green West, Cuffe Lane and Suffolk Street.
See also Curtis & Co.

DOWLING, Robert
1830-1831 Optician, 21 Lower Sackville Street.
1832 Optician, 8 George's Quay.

DUNCAN, Thomas S.
1859 Jeweller and Optician, 35 Wellington Quay.
1860-1861 Working Optician, 29 Wellington Quay.

1862-1865 Optician, 26 Wellington Quay.
1866-1877 Optician, 29 Wellington Quay.
1878-1881 Optician, 20 Wellington Quay.
Duncan's Christian name is given as Thomas only in the 1859, 1880 and 1881 directories. It is possible that more than one man is involved.

EDWARDS, Augustin William
1875-1879 Optician, 27½ Wellington Quay.
1883-1897 Optician, 28 Wellington Quay.
From 1875 to 1879 Edwards' first name is given as Augustus, and he does not appear in the directories for 1880 to 1882. It is possible that more than one man is involved.

EDWARDS, E.
1872-1874 Optician, 27½ Wellington Quay.

FANNON, Andrew
1866-1867 Optician, 31 Essex Street East.

FAWCETT, John
1775-1786 Watchmaker and Optician, 34 Dame Street.
1787-1792 Watchmaker and Optician, 3 Grafton Street.
1793 Watchmaker and Optician, 133 Capel Street.
May have been related to the following reported death "On Essex-quay Mrs Ann Fawcett, relict of the late William Fawcett, Optician", *Freeman's Journal*, 27 February 1768.

FERONI, Alexander
1809-1810 Barometer Maker, 49 Fleet Street.
In 1809 his name is spelt "Fevony". Westropp lists Alexander Feron (sic), mathematical instrument maker, 1781[15].

FLANAGAN, Patrick
1824 Mathematical Instrument Maker, 13 Fishamble Street.
Appears only in Pigot & Co.'s *City of Dublin and Hibernian Provincial Directory*, London, 1824; does not appear in Wilson's *Dublin Directory* for either 1824 or 1825.

FOWLER, John
1759 Mathematical Instrument Maker.
Referred to by Westropp[16].

GATTY, Andrew I
1796-1798 Barometer and Thermometer Maker, 18 Fishamble Street.
1799-1800 Barometer and Thermometer Maker, 1 Smock Alley.

GATTY, Andrew II
1815-1819 Barometer and Thermometer Maker, 1 Smock Alley.
1820-1822 Barometer and Thermometer Maker, 45 Fishamble Street.
1823-1824 Barometer and Thermometer Maker, 7 Smock Alley.

GATTY, Joseph
1801-1807 Barometer and Thermometer Maker, 1 Smock Alley.
1808 Barometer and Thermometer Maker, 25 Fishamble Street.
1809-1814 Barometer and Thermometer Maker, 1 Smock Alley.

GLAZIER, James
1796-1799 Mathematical Instrument Maker, 59 Townsend Street.

GOODWIN, William
Spectacle Maker.
Died 1662[17].

GREGORY, Samuel
1761-1762 Barometer Maker, Sycamore Alley.

GRUBB, Henry G., & Co.
1854-1857 Manufacturers of Philosophical Instruments, Metal Billiard Tables, etc., Charlemont Bridge Works.

GRUBB, Henry S.
1855-1861 Philosophical Instrument Manufacturer, Grand Canal Bank, Rathmines.

GRUBB, Howard (1844-1931)
1873-1874 Optical and Mechanical Engineer, 17 Leinster Square, Rathmines.
1875 Manufacturer of Astronomical Instruments, Observatory Lane.
1876 FRAS, Manufacturer of Astronomical Instruments, 57 Rathmines Road.
1877 FRAS, Astronomical and Mechanical Engineer, 57 Rathmines Road.
1878-1887 FRAS, Astronomical Instrument Manufacturer, 57 Rathmines Road.
1888-1889 Sir Howard Grubb FRAS, Astronomical Instrument Manufacturer, 57 Rathmines Road.
1890-1903 Sir Howard Grubb FRS, Astronomical Instrument Manufacturer, 57 Rathmines Road.
1904 Sir Howard Grubb FRS, FRAS, Astronomical Instrument Manufacturer, 57 Rathmines Road.
1905-1913 Sir Howard Grubb FRS, FRAS, Astronomical Instrument Manufacturer, 54c and 57 Rathmines Road.
1914 Sir Howard Grubb FRS, FRAS, Astronomical Instrument Manufacturer, 54c and 57 Rathmines Road, and Charlemont Bridge Works.
1915-1921 Sir Howard Grubb FRS, FRAS, Astronomical Instrument Manufacturer, 57 Rathmines Road and Charlemont Bridge Works.

GRUBB, Sir Howard & Sons Ltd.
1915-1921 Astronomical Instrument Manufacturer, 57 Rathmines Road and Charlemont Bridge Works.

GRUBB, Thomas (1800-1878)
1838-1841 Engineer and Optician, 1 Upper Charlemont Street.
1842-1844 Engineer to the Bank of Ireland, 1 Upper Charlemont Street.
1845-1848 Optician and Machinist, 1 Upper Charlemont Street.
1849-1854 Engineer to the Bank of Ireland, 1 Upper Charlemont Street.
1855-1858 Engineer to the Bank of Ireland, 14 Leinster Terrace, Rathmines.
1859-1863 Engineer to the Bank of Ireland, 15 Leinster Square, Rathmines.
1864-1878 Engineer to the Bank of Ireland, 141 Leinster Road, Rathmines.

HARRIS, Daniel
1848-1849 Optician, 6 Bonnavilla, Mount Pleasant.
1850-1852 Optician, 4 Bonnavilla, Mount Pleasant.

HARRIS, Daniel and William

1847	Opticians, 8 Wicklow Street.
1848	Opticians, Upper Richmond Street.

HARRIS, Philip & Co. (Ltd)

1902-1904	Laboratory Furnishers and Physical Apparatus Makers, 179 Great Brunswick Street.
1904-1906	Manufacturers of Physical Apparatus, 179 Great Brunswick Street.
1906-1908	Manufacturers of Physical Apparatus, 179 and 182 Great Brunswick Street.
1908-1911	Manufacturers of Chemical and Physical Apparatus, 182 Great Brunswick Street.
	This was a branch of the Birmingham firm Philip Harris & Co. (Ltd).

HEALY, John

1806-1812	Electrifying Machine Maker, 43 James's Street.

HEALY, Samuel

1795-1805	Electrifying Machine Maker, 43 James's Street.
1813-1830	Electrifying Machine Maker, 43 James's Street.
1831	Electrician, 43 James's Street.
1833-1839	Mathematical Instrument Maker, 43 James's Street.
	Healy does not appear in the 1832 directory. A "Mr Healy, optician" married a Mrs Tarvers, widow of Francis Tarvers of Dublin in June 1778[18].

HICKS, George

1824	Mathematical Instrument Maker, 64 City Quay.
	Appears only in Pigot & Co.'s *City of Dublin and Hibernian provincial Directory*, London, 1824; does not appear in Wilson's *Dublin directory* for either 1824 or 1825.
1835	Mathematical Instrument Maker, 68 Marrowbone Lane.

HICKS, J.

1817-1819	Mathematical Instrument Maker, 11 Camden Street.
1820-1822	Mathematical Instrument Maker, 13 Lower Camden Street.
1823-1824	Mathematical Instrument Maker, 13 Peter's Place, Lower Camden Street.
1825-1826	Mathematical Instrument Maker, 64 City Quay.

HICKS, John

1832	Mathematical Instrument Maker, 48 City Quay.
1833-1834	Mathematical Instrument Maker and Circulating Library, 48 City Quay.
1835-1837	Mathematical Instrument Maker and Circulating Library, 20 City Quay.
1838-1839	Mathematical Instrument Maker and Library, 20 City Quay.
1840	Nautical Instrument Maker and Stationer, 20 City Quay.
1841-1843	Mathematical and Nautical Instrument Maker and Stationer, 20 City Quay.
1844	Mathematical and Nautical Instrument Maker, 20 City Quay.
1845-1846	Mathematical Instrument Maker, 20 City Quay.
1847-1848	Mathematical and Nautical Instrument Maker, 20 City Quay.
	In 1824 Pigot & Co.'s *City of Dublin and Hibernian provincial directory* gives Hicks' initial as "P". In 1834 the GPO *Directory* mistakenly gives Hicks' name as Hickman.

HICKS, Thomas

1784-1790 Mathematical Instrument Maker, 40 Poolbeg Street.
1791-1799 Mathematical Instrument Maker, 20 Hawkins' Street.
From 1784-1789 the name appears as "Hicks, T."

HODDER, George

1711 Mathematical Instrument Maker, Essex Street.
Referred to by Westropp[19].

HODGES & SMITH

1858 Grafton Street.
At University College, Galway is a globe marked "Malby's Terrestrial Globe . . . Hodges & Smith Agents Grafton Street Dublin . . . Jany 1858".

JACKSON, Anne and James

1747-1763 Glass Grinders, 5 Essex Bridge.
1764-1774 Glass Sellers, 5 Essex Bridge.
The Jacksons are known to have been trading in 1747 from a bill[20]. They first appear as "glass grinders" in the 1761 directory, in succession to Thomas Jackson[21]. They sold optical instruments.

JACKSON, James and Richard

1775-1785 Glass Sellers, 5 Essex Bridge.

JACKSON, Richard

1786-1827 Carver and Gilder and Looking-glass Warehouse, 5 Essex Bridge.

JACKSON, Thomas

1748 Glass Seller, Essex-Bridge.
He is known, from an advertisement, to have sold optical instruments[22].

JONES, John

1840-1842 Optician in General, 28 Wellington Quay.
1843-1844 Optician in General, 29 Wellington Quay.
1845 Optician, 28 Wellington Quay.
1846-1847 Optician, 29 Wellington Quay.
1848-1854 Optician, 28 Wellington Quay.
1855-1859 Optician, 29 Wellington Quay.
1860 Optician, 28 Wellington Quay.

JONES, Nicholas J.

1860-1861 Optician, 29 Wellington Quay.
Successor to John Jones. In the 1860 directory the name is given merely as Nicholas Jones.

KEARY, John F.

1908-1909 Optician, 13 Wellington Quay.

KEARY, Patrick J.

1886-1889 Optician, 17½ Wellington Quay.
1890 Philosophical Instrument Maker, 17½ Wellington Quay.
There seems to be a close connection between this business and that of P. Cahill, who appears to have shared the same premises.

KEARY, Mrs
1891-1895 Optician, 13 Wellington Quay.

KEARY, P.J.
1892 Optician, 13 Wellington Quay.
1896-1899 Optician, 13 Wellington Quay.
1908-1909 Optician, 13 Wellington Quay.

KEARY, Patrick Cahill
1900-1907 Consulting Optician and Philosophical Instrument Maker, 13 Wellington Quay.

KING, William
1767-1772 Mathematical Instrument Maker, Temple Bar.
A large mahogany octant with boxwood scale, divided diagonally, is stamped on the scale "Made by Willm King Dublin for Mr Iohn Iackson 1784"; private collection: information from Peter Delehar. William King married a Miss Miller in 1764[23]; Westropp states that he died in 1771 but a newspaper announcement provides a different Christian name[24]. King claimed to have been trained by Thomas Cave as his only pupil[25], despite an earlier, similar claim by Jonathan Barker[26].

LAMBERT, Michael
1881 Optician, 47 Rathmines Road.
Appears only in *Slater's Royal National Commercial Directory of Ireland*, Manchester, 1881.

LEVEY, J.
1833-1835 Jeweller and Optician, 39 Mary Street.

LEWIS, John
fl. 1679-1693 Dublin.
Fennell cites him as being active as a watchmaker[27]. A circumferentor signed "Johannes Lewis Dublini fecit anno Domini 1688" is at Maynooth College. Christie's sold a brass sundial signed "Johans Lewis, Dublini fecit, latt 53d 20m, anno 1684" on 12 October 1965 Lot 7. Westropp notes a circumferentor signed "Johann Lewis Dublini Fecit 1693"[28].

LORT, John
1767 Mathematical Instrument Maker, Walker's Alley[29].
1776 Mathematical Instrument Maker, 8 Upper Blind Quay.
1777-1778 Mathematical Instrument Maker, 8 Exchange Street.
1779-1782 Mathematical Instrument Maker, Plunket Street.

LYNCH, James I
1767-1772 Dublin.
Known from a bill in Trinity College, Dublin (see Chapter Three). May have been in business with Edward Spicer, as the title-page of John Hood, *Tables of difference of latitude and departure for navigators, land surveyors, &c.*, Dublin, 1772, advertises "also Sold by EDWARD SPICER and JAMES LYNCH, Mathematicail [sic] Instrument Makers; where may be had the chequered Paper and Compass Theodolite, mentioned in this Book, and all other Instruments for Surveyors, navigators, &c." The same advertisement also appeared in the *Hibernian Journal*, 27 May 1772.

LYNCH, James II

1784-1807 Optician, 26 Capel Street.

LYNCH, James II and III

1826-1831 Optician, 26 Capel Street.

1832-1839 Optician, Mathematical and Philosophical Instrument Maker to the Ordnance and Trinity College, 26 Capel Street.

James Lynch II died in 1833 and was succeeded by James Lynch III (see Chapter Three).

LYNCH, James & Co.

1845-1846 Opticians, 26 Capel Street.

LYNCH, James III and George

1840 and 1843-1844 Opticians, Mathematical and Philosophical Instrument Makers to the University and Royal Dublin Society, 26 Capel Street.

LYNCH & Co.

1841-1842 Opticians to the University and Royal Dublin Society, 36 Westmorland Street.

LYNCH, James II & Son

1808-1825 Opticians, 26 Capel Street.

Sotheby's sold a pocket globe signed "New Terrestrial Globe by Lynch & Son Dublin 1813" on 27 May 1982 Lot 10, and on 23 October 1985 Lot 329 a bronze horizontal sundial signed "Lynch & Son Dublin M.K. 1824".

McEVOY, George

1743 Dublin.

A backstaff, signed "Made by Geo. McEvoy, Temple Bar, Dublin for John Pearson 1743" has been recorded in a private collection.

McGIVENY, Miss

1903-1904 Cutler, Optician and Surgical Instrument Maker, 41 Henry Street.

McGIVENY, Patrick

1900 Optician and Surgical Appliance Maker, 41 Henry Street, and 5 Aston's Quay.

1901 Cutler, Optician and Surgical Appliance Maker, 41 Henry Street and 5 Aston's Quay.

1902-1916 Cutler, Optician and Surgical Instrument Maker, 41 Henry Street.

1914-1922 Cutler, Optician, 41 Henry Street.

From 1900 to 1904 his name is spelt McGiveny, between 1905 and 1913 spelt McGiveney and from 1914 to 1922 McGivney. In 1917 his address is given as 12 Moore Street.

McMENAMY, S.

1815-1816 Military and Naval Mathematician and Ship Chandler, 93 Rogerson's Quay.

1817-1824 Ship Chandler, 93 Rogerson's Quay.

McMENAMY, W.

1810-1814 Military and Naval Mathematician and Ship Chandler, 93 Rogerson's Quay.

1815 Ship Chandler, 11 Poolbeg Street.

This is probably William McMenamy, Master of the Marine School, Dublin, and author of A new method of calculating the area of a survey universally, *Transactions of the Dublin Society*, 1, 1800, 3-10.

McQUAY, William

1815-1817 Mathematical Instrument Maker, 29 Great Strand Street.
1818-1820 Mathematical Instrument Maker, Bridgefoot Street.

MAIBEN, J.M.

1913-1916 Laboratory and Assay Furnisher, 11 Westland Row.
1917-1918 Laboratory Furnishers, 11 Westland Row.

MAIBEN, J.M. & Co.

1912 Laboratory Furnishers, 11 Westland Row.
1919-1922 Laboratory Furnishers, 11 Westland Row.

MARGAS, John (d. 1767)

1761-1767 Optician, Capel Street.
Clay and Court record a sundial marked "MARGAS, Dublin, Made for T. O'Connor, 1767"[30]. Margas had been an apprentice of Nathaniel Adams; his own business in Rose Street, St Martin in the Fields, London, failed in 1758[31]. He was making instruments in Dublin by September 1759[32]; his death was announced in late 1767[33].

MASON, George

1880-1884 Optician, 22 Wellington Quay.
1885-1886 Optician, 24 Wellington Quay.
1887 Optician, 21 Wellington Quay.
1888-1892 Optician, 24 Wellington Quay.

MASON, Jonathan (1784-1849)
MASON, Thomas I (1781-1837)
MASON, Thomas and Jonathan

1805-1808 Mason, Thomas and Jonathan, Opticians, 8 Arran Quay.
1809 Mason, Jonathan, Optician, 8 Arran Quay.
1809 Mason, Thomas, Optician, 4 Essex Bridge.
1810-1812 Mason, Jonathan, Optician, 9 Ormond Quay.
1810-1812 Mason, Thomas, Optician, 3 Essex Bridge.
1813-1817 Mason, Thomas and Jonathan, Opticians, 3 Essex Bridge.
1818-1822 Mason, Jonathan, Optician, 14 Capel Street.
1818-1819 Mason, T., Optician, 3 Essex Bridge.
1820 Mason, T., Optician, 4 Essex Bridge.
1821-1826 Mason, T., Optician, 3 Essex Bridge.
1827-1833 Mason, T., Optician, 6 Essex Bridge.
1834-1836 Mason, Thomas, Optician, 6 Essex Bridge.
1837-1838 Mason, Thomas, Optician to His Excellency the Lord Lieutenant, 6 Essex Bridge.
Successors to Seacombe Mason I. Jonathan Mason moved to Limerick after his second marriage to a Limerick woman[34].

MASON, Seacombe or Seacome I (1745-1804)

1780-1804 Optician, 8 Arran Quay.
From 1780-1783 Mason's Christian name is spelt "Seacom", in the directory and from 1784-1787 spelt "Seacome". In 1783 the address is given merely as "Arran Quay".

MASON, Seacombe or Seacome II (1808-1892).

1838	Optician and Mathematical Instrument Maker, 6 Essex Bridge.
1839-1844	Optician to His Excellency the Lord Lieutenant, 6 Essex Bridge.
1845-1846	Optician, 11 Essex Bridge.
1847-1864	Optician to the Lord Lieutenant, 11 Essex Bridge. Successor to Thomas Mason I, succeeded by Mason & Son.
1877	Optician, 11 Essex Bridge.
1878	Optician, 11 Upper Ormond Quay. He is normally entered merely as Seacombe or Seacome, but the 1878 directory lists him as "Mason, Seacombe, sen."

MASON, Seacome III

1877	State Optician, 11 Nassau Street.
1879	Optician, 24 Wellington Quay. In the 1877 directory Mason is listed as "Mason, Seacombe, jun.", he is not listed in 1878, and in 1879 he is "Mason, Seacombe".

MASON, Standish

1839	Optician, 2 Upper Ormond Quay.
1840-1841	Optician and Mathematical Instrument Maker, 2 Upper Ormond Quay.

MASON, Thomas I *see* MASON, Jonathan

MASON, Thomas II (1840-1913)

1866-1868	Optician, 9 Nassau Street.
1878-1883	Optician, 11 Essex Bridge.
1884-1895	Optician, 21 Parliament Street.
1896-1899	Optician, 4 Dame Street.
1900-1916	Optician, 5 Dame Street.
1917-1922	Optician, 5 and 6 Dame Street.

MASON, Thomas and Jonathan *see* MASON, Jonathan

MASON & Co.

1876	Opticians to the Lord Lieutenant, 11 Essex Bridge. Successors to Mason & Son, succeeded by Seacombe Mason II. A stick barometer signed "Mason & Co. 11 ESSEX BRIDGE Dublin PRESENTED TO THE Grand Lodge of Ireland by SEACOME MASON JUNR. P.M. LODGE 93. AUGUST. 1875", has been recorded by Dr Charles Mollan.

MASON & Son

1865-1875	Opticians to the Lord Lieutenant, 11 Essex Bridge. Succeeded by Mason & Co.

MELESCO, Dominick *see* MILESSIO, Dominick

MILESSIO, Dominick

1812	Carver and Gilder, 25 High Street.
1813-1819	Carver and Gilder, 35 Skinner Row.
1820-1824	Looking-glass Manufacturer, 6 Skinner Row.
1825	Looking-glass Manufacturer, 11 Skinner Row. In 1812 the surname is given as "Melesco" and "Milessio". There is an 8-inch wheel barometer, signed "D. Milefsio DUBLIN" in the National Museum of Ireland; another, signed "C. Milesio Fecit" was sold at Sotheby's

2 December 1977 Lot 36; and a third is recorded by Goodison, who also notes an example by "D. Milesio" of Belfast[35].

MONAGHAN, James
1760-1762 Weatherglass Maker, Winetavern Street.

MOORE, F.M.
1864 Watchmaker and Optician, 23 Eden Quay.
1865 Watchmaker and Optician, 23 Eden Quay and Belfast.
1866-1900 Watchmakers and Opticians etc., 23 Eden Quay and Belfast.
1901 F.M. Moore Ltd, Watchmakers and Nautical Instrument Manufacturers, 23 Eden Quay and Belfast.
In the directories for 1864 to 1866 the name is listed as "F.M.M. Moore".

MOORE, Lewis
1867 Optician and Lampist, 4 King Street South.

MORROW, Thomas
1902-1903 Manufacturer of Scientific Instruments, Calcium Carbide, etc., 31 Lincoln Place.

MULLALLY, Luke
1751 Maker of Barometers and Mercurial Thermometers, Capel-street[36].

NELSON, William
1830-1831 Spectacle Maker and Jeweller, 21 Essex Quay.
1832 Optician and Jeweller, 20 Essex Quay.
1833 Optician and Jeweller, 24 Essex Quay.
1834-1836 Optician and Jeweller, 37 Lower Ormond Quay.
1837-1839 Optician, 37 Lower Ormond Quay.
Nelson does not appear in the 1838 Directory.

NELSON, William H.
1840-1844 Optician, 37 Lower Ormond Quay.
1845-1851 Optician, 42 Lower Ormond Quay.
1852-1860 Optician and Mathematical Instrument Maker, 66 Dame Street.
1861-1862 Optician and Mathematical Instrument Maker to the Lord Lieutenant, 66 Dame Street.
This may be the same man as William Nelson.

NUEL, E.
1867 Watchmaker and Optician, 33 Eden Quay.

PARKER, George
1761-1766 Watchmaker, Cork Hill.
The George Parker who voted at the election 1761 was a Goldsmith[37]. There is an equatorial ring dial signed "GEORGE PARKER – DUBLIN" in a private collection.

PERRY, Andrew
1839 Optician, 5 Fitzwilliam Place, Grangegorman Lane.
1840-1842 Mathematical Instrument Maker, 5 Fitzwilliam Place, Grangegorman Lane.
1843-1845 [no trade], 21 Grangegorman Lane.

PERRY, Thomas
1835-1837 Optician, 22 Grangegorman Lane.

PITZOLI, Anthony
1831 Weatherglass Maker, 86 Pill Lane.
1832 Weatherglass Maker, 86½ Pill Lane.
1833-1835 Weatherglass Manufacturer, 86½ Pill Lane.

PRESCOTT, George & Co.
1879-1882 Philosophical Instrument Makers, Watchmakers and Opticians, 8 South King Street.

PRESCOTT, George & Co. Ltd
1883-1884 Electrical Engineers and Philosophical Instrument Makers, 9 Merrion Row.

PRESCOTT, George
1895-1901 Electrical Engineer and Philosophical Instrument Maker, 9 Merrion Row.
1905 Electrical Engineer, 9 Merrion Row.

PRESCOTT, George Ltd
1896 Manufacturing Optician, 1 Duke Street and Edinburgh.

PRESCOTT, George & Sons
1902-1904 Electrical Engineers and Philosophical Instrument Makers, 9 Merrion Row.
1906-1911 Opticians and Electrical Engineers, 9 Merrion Row.
1912-1922 Opticians and Electrical Engineers, 9 Merrion Row and 42 Wellington Place, Belfast.

R., W.
fl. 1667 Dublin.
The Museum of the History of Science, Oxford, has a circumferentor signed "W.R. Dublin. fe. latitude 53° 20′ *1667*".

RICHARDSON, William
1784-1786 Optician, 4 East Cole Alley.

ROBINSON, James
1845 Museum of Curiosities, 41 Grafton Street.
1846-1851 Philosophical Artist and Optician, Polytechnic Museum and Gallery of Curiosities, 65 Grafton Street.
1852-1855 Optician, Polytechnic Museum and Gallery of Curiosities, 65 Grafton Street.
1856-1884 Optician, Polytechnic Museum and Photographic Gallery, 65 Grafton Street.

ROBINSON, James & Sons
1885-1886 Optician, Polytechnic Museum, Photographic Gallery, etc., 65 Grafton Street, and 172 Regent Street, London.
1887 Optician, Polytechnic Museum, Photographic Gallery, etc., 65 Grafton Street.
1888 Optician, Polytechnic Museum, Photographic Gallery, etc., 65 Grafton Street, (temporary premises) 23 Nassau Street.
1889-1899 Optician, Polytechnic Museum, Photographic Gallery, etc., 65 Grafton Street.
1900-1901 Photographic Gallery, etc., 65 Grafton Street.
1902-1903 Photographic Gallery and Photographic Supply Merchants, 65 Grafton Street; Photographic Supply Merchants, 37 College Green.

ROBINSON, J. & Sons Ltd

1904-1910 Photographic Gallery and Photographic Supply Merchants &c., 65 Grafton Street.

ROBINSON, John

1791-1792 Mathematical Instrument Maker, 4 Drogheda Street.

1794-1795 Mathematical Instrument Maker, 21 Hawkin's Street.

Robinson does not appear in the 1793 directory.

ROPER, Abraham

1787-1788 Abraham Rosser, Mathematical Instrument Maker, 8 Bedford Row.

1789-1793 Abraham Roper, Mathematical Instrument Maker, 8 Bedford Row.

It is almost certain that these two are the same man.

ROSSER, Abraham *see* ROPER, Abraham

SAMUELS & CO.

1867-1868 Opticians, 29 Nassau Street.

SANDHEIM, B.J.

1876 Scientific and Practical Optician, 30 Nassau Street.

SAUNDERS, M.A.

1821-1822 Optician, Eden Quay.

1823-1829 Optician and Bristol Packet Office, 27 Eden Quay.

1830-1831 Optician, 6 Aston Quay.

SAUNDERS, Thomas

1793-1794 Mathematical Instrument Maker, 7 George's Street.

1796-1799 Optician, 35 College Green.

1800-1803 Optician, 6 Church Lane.

1804-1818 Optician, 6 S.E. Church Lane.

1819 Optician, Eden Quay.

Saunders does not appear in the 1795 directory. Sotheby's sold a portable sundial by Saunders of Dublin, the lid with a silver label "I. Burgh, Esq.," the inside of the lid with a paper label with compass variations and the date June 1794, on 1 November 1965 Lot 70. A "Mr Saunders, optician etc., George Street" married a Miss Fitzgibbon of Youghal in September 1793[38].

SCOTT, David

1825-1838 Mathematical Instrument Maker, 14 Stafford Street.

SEWARD, –

c.1750 Dublin.

Name, address and dates so far untraced. Known from a number of instruments[39].

SIMONTON, James

1863 Photographer etc., and Optician, Royal Panopticon of Science and Art, 70 Grafton Street.

1864-1876 Photographer and Optician, Royal Panopticon of Science and Art, 70 Grafton Street.

SMYTH, Victor E.

1899-1902 Optician and Photographic Dealer, 8 D'Olier Street.

1903 Optician and Photographic Depot, 8 D'Olier Street.
1904 Optician and Photographic Depot, 6 Westland Row.
1905-1910 Optician and Photographic Depot, 6 Westmoreland Street.

SOLOMONS, M.E.
1856-1859 Optician, 19 Nassau Street.
1860 Optician to the Royal Family, 19 Nassau Street.
1861-1899 Opticians to the Royal Family and Patentee of Spectacles, 19 Nassau Street.
1900-1905 Optician, 19 Nassau Street.

SPEAR, Richard I
1791-1792 Optician, 29 Capel Street.
1793-1795 Optician, 23 Capel Street.
1796-1808 Mathematical Instrument Maker, 23 Capel Street.
1809 Mathematical Instrument Maker, 23 Capel Street and College Green.
1810 Mathematical Instrument Maker, 35 College Green.
1812-1814 Mathematical Instrument Maker, 27 College Green.
In 1791 and 1792 his surname is given as "Speer".

SPEAR, Richard II (?)
1818 Mathematical Instrument Maker, and Oil Merchants, Wax and Spermaceti-candle Manufacturers, 27 College Green.
1819-1835 Mathematical Instrument Maker, 27 College Green.
1836-1837 Optician, Wax and Oil Merchant, 27 College Green.
In 1818 the directory entry is "Spear, Richard & Co., Mathematical Instrument Manufacturers". From 1820 to 1831 there is a separate entry for "Spear, Richard & Co., Spermaceti-refiners and Oil Merchants", at the same address.

SPEAR, Saunderson
1831 Wax and Spermaceti-candle Manufacturer, 3 Westmorland Street.
1832-1833 Wax Chandler, 98 Grafton Street.
1834 Wax Chandler, Oil Merchant and Fancy Jeweller, 98 Grafton Street.
1835 Wax Chandler, Oil Merchant and Fancy Jeweller, 23 Suffolk Street.
1836-1837 Optician, Wax Chandler and Oil Merchant, 23 Suffolk Street.
A Saunderson Spear married Ann Hope in 1826[40].

SPEAR & CLARKE
1815-1817 Mathematical Instrument Makers, 27 College Green.
Successors to Richard Spear and Edward Clarke.

SPEARS & Co.
1838 Opticians, Wax and Oil Merchants, 27 College Green.
1839-1840 Opticians and Mathematical Instrument Makers, 27 College Green.
1841 Opticians and Mathematical Instrument Makers to Her Majesty's Revenue in Ireland, Wax and Spermaceti Merchants, 27 College Green.
1842 Opticians and Mathematical Instrument Makers to Her Majesty's Board of Customs, Wax and Spermaceti Merchants, 27 College Green.
1843-1844 Opticians and Mathematical Instrument Makers to Her Majesty's Board of Customs, Wax and Spermaceti Merchants, 28 College Green.
1845-1846 Opticians, 28 College Green.
1847 Opticians, Mathematical Instrument Makers, Wax and Oil Merchants to his Excellency the Earl of Bessborough, 28 College Green.

1848-1850 Opticians and Mathematical Instrument Makers to his Excellency the Earl of Clarendon, Wax Chandlers and Oil Merchants, 28 College Green.
1851 Opticians to his Excellency the Earl of Clarendon, 28 College Green.
1852 Oil Merchants and Opticians to his Excellency the Earl of Clarendon, 28 College Green.
1853-1855 Oil Merchants and Opticians, 28 College Green.
1856-1860 Opticians, 28 College Green.
1861-1864 Opticians to the Queen, Agents for the Sale of Charts and Mathematical Instrument Makers, 28 College Green.

The 1838 GPO *Directory* gives the name of the firm as Spears & Co. Some of the above variations are due to the different directories available: the listings up to 1844 are from the GPO *Directory*, those from 1845 from Thom's.

SPENCER, James
1861-1868 Optician, 47 Cullenswood Avenue.

SPENCER, John
1845-1851 Philosophical Instrument Maker, 3 Aungier Street.
1852-1863 Philosophical Instrument Maker, 13 Aungier Street.

SPENCER, John, & Son
1864-1865 Philosophical Instrument Makers, 13 Aungier Street.
1866-1868 Philosophical Instrument Makers, Opticians and Mechanical Engineers, 13 Aungier Street and 19 Grafton Street.
1869-1872 Philosophical Instrument Makers, Opticians and Mechanical Engineers, 19 Grafton Street.
1873-1883 Opticians and Engineering Instrument Manufacturers to the Board of Works, 19 Grafton Street.
1884 Opticians and Engineering Instrument Manufacturers to the Board of Works, 23 Nassau Street.
1885-1886 State Opticians and Engineering Instrument Manufacturers to the Board of Works, 23 Nassau Street.

SPICER, Edward
fl. 1768-1772 Mathematical Instrument Maker, Plunket Street.

Spicer's instruments are recommended by Benjamin Noble, *Geodaesia Hibernica*, Dublin, 1768, and in turn, "E. Spicer, mathematical Instrument-Maker Plunket-street" sold the book according to an advertisement in *Freeman's Journal*, 5 April 1768. A miner's dial signed "Spicer Dublin" with "John Garman Doneraile 1771" was sold at Sotheby's 5 March 1981 Lot 16. Spicer also advertised "the chequered Paper and Compass Theodolite, mentioned in this Book, and all other Instruments for Surveyors, navigators, &c." in John Hood, *Tables of difference of latitude and departure for navigators and land surveyors, &c. . .*, Dublin, 1772. He may have been in business briefly with James Lynch I. Two Edward Spicers voted in Dublin at the election 1761: one of them must almost certainly have been the instrument maker[41]. Spicer advertised in the *Hibernian Journal*, 27 May 1772. He died around 1775[42]. The International Genealogical Index records four children of Edward and Ann Spicer, all christened in St Michael the Archangel, Dublin: Mary, 10 April 1766; Thomas, 10 March 1767; Andrew, 3 December 1769; and Susanna, 15 March 1772.

STANLEY, William C.

1897-1899	Scientific Instrument Maker, 31A Lincoln Place.
1900	Scientific Instrument Maker, 1 Clare Street and 31 Lincoln Place.
1901	Scientific and Mathematical Instrument Maker, 31 Lincoln Place.

STANLEY, W. Oswald

1898-1902	31A Lincoln Place.
	Appears only in the trades section of the directory, presumably related to and in business with W.C. Stanley.

STANLEY & Co.

1900	Opticians, Electricians etc., 1 Clare Street.

STEPHENS, Alexander

fl. 1747	Dublin.
	A backstaff signed "Made by Alexander Stephens for Gyles Fearon 1747" is in a private collection; another signed "Made by Alex^{r} Stephens Temple Barr DUBLIN For George Waddrington" is in the Whipple Museum, Cambridge, inventory no. 698.

STEPHENS, Alexander and John

1760-1774	Mathematical Instrument Makers, Temple Bar.
1775-1781	Mathematical Instrument Makers, 31 Temple Bar.

STEPHENS, John

1782-1793	Mathematical Instrument Maker, 31 Temple Bar.
	Westropp states that John Stephens died in 1802[43].

STOKES, Gabriel (bapt. 1682-d.1768)

fl. 1715-1742	Essex Street.
	A horary quadrant in the National Maritime Museum, Greenwich, is signed "Gab: Stokes Dublin Fecit 1738", inventory no. Gabb Collection F.17. A 36-inch Gunter wooden sliding rule in the Whipple Museum, Cambridge is signed "Gabriel Stoakes Dublin fecit 1719 For Iohn Connell", inventory no. 177. A horizontal dial in the Willis Museum and Art Gallery, Basingstoke is signed: "Stokes fecit 1742", inventory no. BWM 1958.4.

STOKES, William

1842	Optician, 3 Upper Leeson Street.
1843-1845	[no trade], 3 Upper Leeson Street.
1846-1850	(of Spears & Co.), 3 Ballygihen, Sandycove, Kingstown and 28 College Green.

TERNAN, William G.

1873-1875	Watchmaker, Jeweller, Optician and Bog Oak Manufacturer, 29 Nassau Street.
	In the directories for 1873 and 1874 Ternan's initials are given as G.W.

TICKELL, G.

1827-1828	Optician, 74 Great Britain Street.
1829-1840	Optician, 24 Upper Sackville Street.
	The 1840 entry is taken from the GPO *Directory*.

TICKELL, George

1840 Optician, 48 Beresford Street.

Perhaps the same as the above: the 1840 entry is taken from the Pettigrew and Oulton *Directory*.

TOTTENHAM, Edward

1844 Optician, Watch and Clockmaker (late of Spears and Co.), 38 College Green.

1845-1847 Optician, 38 College Green.

VECCHIO, J. Del *see* Del VECCHIO, J.

WALKER, Frederick

1832-1833 Mathematical Instrument Maker, 7 Trinity Place.

1834-1840 Mathematical Instrument Maker, 31 Clarendon Street.

1841-1850 Mathematical Instrument Maker, 29 Clarendon Street.

WALKER, George

1841-1848 Mathematical and Optical Instrument Maker, 6 Fade Street.

WALKER, W.

1820-1826 Mathematical Instrument Maker, 17 Temple Bar.

WALKER, William

1775-1790 Mathematical Instrument Maker, 15 Temple Bar.

1791-1804 Mathematical Instrument Maker, 17 Temple Bar.

At their sale on 2 February 1984 Lot 84 Phillips auctioned a miner's dial signed "Walker Dublin 1792". A circumferentor in a private collection is signed "Walker * Dublin 1797 R[?] Daly[?]". Sotheby's sold a silver-mounted shagreen-covered mathematical instrument case by Walker, with the owner's name Henry O'Connor and the date 1779 inscribed on the combined scale and protractor, at their sale on 27 October 1969 Lot 17.

WALKER, William & Son

1805-1819 Mathematical Instrument Maker, 17 Temple Bar.

The address 16½ Temple Bar is also known from a circumferentor in a private collection. Another circumferentor, also in a private collection, is signed "Walker & Son No.17 Temple Bar DUBLIN 1802", and the National Museum of Ireland has one signed "Walker & Son No.17 Temple Bar DUBLIN Hugh Robinson 1807".

WHITE, William

1824 Optician, 17 Crow Street.

Appears only in Pigot & Co.'s *City of Dublin and Hibernian provincial directory*, London, 1824; does not appear in Wilson's *Dublin directory* for either 1824 or 1825.

WINTER, Thomas

d. 1719 Mathematical Instrument Maker.

"Died 1719"[44]. The International Genealogical Index records Thomas Winter, son of Christopher and Mary Winter, christened on 7 February 1694 at St Michan, Dublin.

WOODSIDE, James

fl. 1733 Mathematical Instrument Maker.

Died 1743[45]. A backstaff dated 1733 survives[46]. The International Genealogical Index records the marriage between James Woodside and Jane Cordner on 7 December 1731 at St Andrews, Dublin.

YEATES, George (1796-1882)

1826-1827	Optician, 70 Camden Street.
1828-1837	Optician, 70 Charlemont Street.
1843-1851	Optician, 2 Grafton Street.
1852-1858	Optician and Mathematical Instrument Maker to the University, 2 Grafton Street.

George Yeates does not appear in the directories for 1835 or 1836. It is not certain that the George Yeates who traded before 1837 is the one who traded after 1843. The business continued after 1858 as George Yeates & Son.

YEATES, George & Son

1840-1846	Opticians, 2 Grafton Street.
1859-1864	Opticians and Mathematical Instrument Makers to the University, 2 Grafton Street.

Successors to George Yeates, and succeeded by Yeates & Son.

YEATES, Horatio (1834-1906)

1859	[no trade], 29 Wicklow Street.
1860-1862	Optician, 29 Wicklow Street.
1863	Optician, etc., 23 Dawson Street.
1864	Optician, etc., 13 Leinster Street.

YEATES, Kendrick (1776-1850)

1803-1811	Optician, 22 Henry Street.
1816	Optician, 27 Stafford Street.
1817-1818	Optician, 26 Stafford Street.
1819-1826	Optician and General Brass Manufacturer, 26 Stafford Street.
1827-1835	Brassfounder and General Brass Merchant, 26 Stafford Street.
1836	Brassfounder and General Brass Merchant, 24 Suffolk Street.
1837-1838	Brassfounder and General Brass Merchant, 26 Stafford Street.
1839	Brassfounder and General Brass Merchant, 33 Stafford Street.

He does not appear in the Directories for 1812 to 1815. In 1818 his name appears as "Kenrick". A Kendrick Yeates married Margaret Faulkner in 1802[47].

YEATES, Samuel (1762-1834)

1790-1794	Optician, 4 Upper Ormond Quay.
1795-1810	Optician, 29 Capel Street.
1811-1826	Optician, 89 Dame Street.
1827-1831	Optician, 2 Grafton Street.

YEATES, Samuel & Son

1832-1838	Opticians, 2 Grafton Street.
1839	Opticians, 2 Grafton Street and 9 Nassau Street.

YEATES, Stephen M.

1865-1902	Optician, 2 Grafton Street.

YEATES, Thomas

1848	Optician, 119 Abbey Street.

YEATES, William

1827	Optician and Cutler, 18 Capel Street.
1828	Optician, 18 Capel Street.

YEATES & Son

1834-1839	2 Grafton Street. Their trade card is illustrated by Crawforth[48].

YEATES & Son

1865	Opticians and Mathematical Instrument Makers to the University, 2 Grafton Street.
1866-1872	Opticians and Mathematical Instrument Makers to the University and Port of Dublin, 2 Grafton Street.
1873-1874	Opticians and Mathematical Instrument Makers to the University and Port and Docks' Board, 2 Grafton Street.
1875-1879	Opticians and Mathematical Instrument Makers to the University and Port and Docks' Board, 2 Grafton Street and 136 St Stephen's Green West.
1880-1891	Opticians and Mathematical Instrument Makers to the University and Port and Docks' Board, 2 Grafton Street.
1892-1902	Opticians and Mathematical Instrument Makers to the University and Port and Docks' Board, 2 Grafton Street; Workshops and Depot, 25 Suffolk Street.
1902-1910	Opticians and Mathematical Instrument Makers to the University and Port and Docks' Board, 2 Grafton Street.
1911-1913	Manufacturing Opticians, 2 Grafton Street and 33 Nassau Street.
1914	Manufacturing Opticians, 2 Grafton Street, 51 Nassau Street and 32A Dawson Street.
1915-1922	Manufacturing Opticians, 2 Grafton Street and 51 Nassau Street.

References and Notes for Chapter Seven

1. For a discussion of Irish directories, see R. ffolliott and D.F. Begley, Guide to Irish Directories, in D.F. Begley (ed.), *Irish genealogy: a record finder*, Dublin, 1981, 75-106. See also Edward Evans, *Historical and bibliographical account of almanacks, directories etc . . .*, facsimile edition, Dublin, 1976, 124-142. An assessment of directories as a primary source is given by M.A. Crawforth, Makers and Dates, *Bulletin of the Scientific Instrument Society*, No. 13 (1987), 2-3.
2. Thomas H. Mason, Dublin opticians and instrument makers, *Dublin Historical Record*, **6**, 1944, 140.
3. Discussed by Joseph W. Hammon, George's Quay and Rogerson's Quay in the eighteenth century, *Dublin Historical Record*, 5, 1942-43, 53-54.
4. M.D. Westropp, "List of Dublin opticians and mathematical instrument makers", manuscript at the Museum of the History of Science, Oxford.
5. Noted in J.H. Andrews, *Plantation acres: an historical study of the Irish land surveyor and his maps*, Belfast, 1985, 302, n. 25, an advertisement placed in the *Dublin Mercury*, 28 and 31 March 1767, which stated: "John Alment, Mathematical Instrument Maker, next door to the Sign of the White Hart, Mary's-abbey, who for several years worked for Mr Margas of Capel-street, but now on his own account, makes the following articles viz. Cases of drawing instruments, theodolites, sircumfronters [sic], sundials, electrical machines, weather glasses, and various other instruments too tedious to insert, all of which he is determined to sell as low as possible, and hopes that, and the goodness of his work, will recommend him. N.B. As said Alment is a new beginner, he hopes for encouragement, as it will always be his study to excell in the workmanship of his instruments".
6. An advertisement placed in the *Dublin Mercury*, 6 February 1768, and which reappeared throughout March, April, May and June, stated: "John Alment, Instrument-maker, who for several years worked as foreman to the late Mr Margas

of Capel Street, now carries on business for himself in Mary's-abbey, and has ready for sale, all sorts of surveying instruments, ditto for drawing, curiously made, various sundials, pocket compasses, barometers, electoral [sic] machines, linen provers, with various other instruments". The same newspaper carried a different version of this advertisement on 15, 17 and 20 December 1768: "The improved Barometer, which has had the approbation of the Dublin Society, is made only by John Alment, Optician in Mary's abbey, who manufactured all sorts of optical, philosophical, and mathematical instruments, and as said Alment was for several years foreman to the late Mr Margas of Caple-street [sic], whose children have now entirely thrown off that business, he humbly intreats the favours of his customers, as it will always be his study to merit their countenance and protection."

7. Mason, *op. cit.*, 149 (ref. 2).
8. Westropp, *op. cit.*, (ref. 4).
9. The following advertisement appeared in *Pue's Occurrences*, 12 June 1756: "STOPT, by John Bate in Nicholas-street, an Horizontal Sun Dial, nine Inches square. Whoever can prove the Property may have it by applying to said Bate, and paying the Cost and Charge of Advertising."
10. The Knight of Glin, Dublin directories and trade labels, *Furniture History*, **21**, 1985, 258-282, illustrates their trade card (figure 3). See also his A family of looking-glass merchants, *Country Life*, **149**, 1971, 195-199.
11. *Appendix to the thirtieth report of the deputy keeper of the public records and keeper of the state papers in Ireland: An index to the act or grant books and original wills of the diocese of Dublin from 1800 to 1858*, Dublin, 1899, 124.
12. J.M. Andrews, *op. cit.*, 130 and 291, n.29 (ref. 5), notes an advertisement placed in the *Dublin Journal*, 23 May, 10 June 1749. This states: "Jonathan Barker, who served his Apprenticeship for some Years to the late Mr Thomas Cave, Land Surveyor and Mathematical Instrument Maker, intends to follow the same business as he is the only Person to whom Mr Cave taught Surveying ...". Westropp, *op. cit.*, (ref. 4), states that Cave died c.1750.
13. See also The Knight of Glin, Dublin directories ..., *op. cit.*, 262-265 (ref. 10).
14. Westropp, *op. cit.*, (ref. 4).
15. *Ibid.*
16. *Ibid.*
17. *Ibid.*
18. Henry Farrar, *Irish marriages: being an index to the marriages in Walker's Hibernian Magazine 1771 to 1812*, London, 1897, 208.
19. Westropp, *op. cit.*, (ref. 4).
20. The Knight of Glin, Dublin directories ..., *op. cit.*, 267 (ref. 10).
21. The following advertisement appeared in the *Dublin Journal* on 2 and 6 January 1759: "ANN Jackson and JAMES JACKSON, Widow and Brother of Thomas Jackson, Glass-seller on Essex Bridge, Intend following the Business in the same extensive Manner as it has been hitherto, and hope to please all those who are so kind as to favour them with their Commands. N.B. The Executors of the said Thomas Jackson have impowered James Jackson to receive all the outstanding Debts; And all those who have any Demands are desired to bring them in, that they may be immediately discharged."
22. An advertisement placed in the *Dublin Journal*, 1 November 1748, and which reappeared throughout December and January 1752 stated: "Thomas Jackson, Glass seller, at the Golden Frame and Spectacles on Essex-bridge, has lately imported a most curious Parcel of Optical Instruments, viz. reflecting Telescopes, on the Principles of Sir Isaac Newton and M. Gregory, made with the greatest Accuracy, refracting Telescopes, Microscopes both single and double, either for the study or the Pocket, the finest Chrystal Spectacles ground on Brass Tools, (approved by the Royal Society) set in all manner of ways that are most convenient and beautiful, Reading Glasses, Opera Glasses, Concave Glasses for short-sighted Persons, Prisms, Sky Optick Balls, Convex and Concave Speculums, Cylindick concaves, Pocket Machines for viewing Pictures, Globes for Halls and Stair Cases, gilt Branches, and Gold Leaf, all at reasonable Prices, either Wholesale or Retail".
23. *The Freeman's Journal*, 1 May 1764.
24. *Hibernian Journal*, 30 December 1771: "Died . . . On Temple-bar, Mr. Thomas King, Mathematical Instrument Maker."
25. Noted in J.H. Andrews, *op. cit.*, 291, n.30 (ref. 5), *Pue's Occurrences*, 29 September 1759.
26. *Dublin Journal*, *op. cit.*, (ref. 12).
27. Geraldine Fennell, *A list of Irish watch and clock makers*, Dublin, 1963, 23.
28. Westropp, *op. cit.*, (ref. 4).
29. An advertisement placed in the *Dublin Mercury*, 3 November 1767 stated: "Whereas John Barry, apprentice to John Lort; mathematical instrument-maker, in Walker's-alley, near St. Patrick's cathedral, Dublin, eloped on Monday evening the 2d instant from his said master's service, without any provocation whatever, this is to inform said Barry, if he will return to said master's service he shall be forgiven; but whoever harbours or employs said Barry, shall be prosecuted as the law directs. Said Barry is about 17 years of age, fresh coloured, wears a wig, and had on when he went away, a claret coloured coat faded, with Bath metal buttons, inlaid with steel, a drab coloured waistcoat and leather breeches. He

humbly hopes that no masters of vessels will take him on board, bound to Liverpool or Chester, as his said master has been creditably informed he intends going to Liverpool. Note: He has 4 years of his time to serve. Dublin November 6th 1767".

30. R.S. Clay & T.H. Court, *The history of the microscope*, London, 1932, 249.
31. E.G.R. Taylor, *The mathematical practitioners of Hanoverian England*, Cambridge, 1966, 268.
32. The following advertisement appeared in the *Dublin Journal* during September and October 1759: "Whereas I, John Mooney of Grashols in the King's County, Land Surveyor, have, from many Years Application, discovered a Geometrical Problem, by which the Longitude at Sea can be ascertained, and Time exactly measured, with many Uses infinitely serviceable to Mankind; from whence, by arithmetical Calculations, any Sailor by the Assistance of the four Rules of vulgar Arithmetick, shall be able to find out the Longitude of any Part of the Globe, as often as the Stars are visible; so that with the utmost Facility and Certainty he will be able to come to the Knowledge of it. And as my Scheme is founded on the unerring Principle of Truth, so my calculations are more certain than any Art (by the Help of Machinery and Instruments) can be without them.
    September 15 1759 JOHN MOONEY
    N.B. I have had made by Mr Margas in Dublin a curious Instrument, which truly represents all the Motions of the Earth round the Sun, and by which the Longitude is illustrated to a Demonstration; and has been lately proved in that Manner by many Gentlemen of Knowledge and Learning, as well in Dublin as in the County".
33. *Freeman's Journal*, 8 December 1767: died "a few Days ago in Capel-street, Mr Margas, Optician".
34. [Thomas H. Mason & Sons Ltd], *The Mason family business: a brief history 1780-1980*, [Dublin, 1980], 4.
35. Nicholas Goodison, *English barometers 1680-1860*, 2nd edn., Woodbridge, 1977, 341.
36. An advertisement placed in the *Dublin Journal*, 13 April 1751, and which reappeared throughout April, May, June and July, stated: "Made by Luke Mullally, at the blue window in Caple-street [sic] near Mary's-street, all Sorts of Jet Necklaces, Pendants, Buttons, Bugles, Glass Toys, Barometers and mercurial Thermometers. As he manufactured the above Goods, and exceeds any imported, will sell by Wholesale or Retail on the most reasonable Terms".
37. H.F. Berry, The Goldsmiths' Company of Dublin, *Journal of the Cork Historical and Archaeoligical Society*, 2nd series, **8**, 1902, 29-50.
38. Farrer, *op. cit.*, 392 (ref. 18).
39. These are a ring dial in a private collection; a mechanical equinoctial dial, inventory no. 578-1910, a circumferentor, inventory no. 84-1940, and a linen prover in the National Museum of Ireland; a pair of brass and steel proportional dividers, offered by Christie's 22 November 1978 Lot 10; and a mechanical universal equinoctial dial offered by Sotheby's 13 November 1961 Lot 78.
40. *Appendix . . ., op. cit.*, 994 (ref. 11).
41. *An alphabetical list of the freemen and freeholders that polled at the election of members to represent the City of Dublin in Parliament*, Dublin, 1761, 67.
42. Westropp, *op. cit.*, (ref. 4).
43. *Ibid.*
44. *Ibid.*
45. *Ibid.*
46. Science Museum, inventory no. 1912-393.
47. *Appendix . . ., op. cit.*, 1151 (ref. 11).
48. M.A. Crawforth, Evidence from trade cards for the scientific instrument industry, *Annals of Science*, **42**, 1985, 541.

CHAPTER EIGHT

PROVINCIAL INSTRUMENT MAKERS

A.D. Morrison-Low and J.E. Burnett

Instrument makers working outside Dublin are listed below. The list is compiled chiefly from directories, listed in chronological order of publication. The following directories were examined:

Richard Lucas, *The Cork directory*, Cork, 1787, reprinted as "The Cork directory for the year 1787", *Journal of the Cork Historical and Archaeological Society*, **72**, 1967, 135-157; and in *The Irish Genealogist*, **4**, 1969, 37-46.
The new Cork directory for the year 1795, Cork, [1795].
Holden's triennial directory . . ., 4th edn., *for 1805, 1806, 1807*, vol. II, London, 1805.
Holden's triennial directory . . ., 5th edn., *for 1809, 1810 and 1811*, vol. II, London, 1809.
William West, *Directory and picture of Cork and its environs*, Cork, 1810.
Connor's Cork directory for the year 1812, Cork, [1812]. Also copies for 1817, 1820, 1826 and 1828.
Thomas Bradshaw, *Directory of Belfast and Lisburn*, Belfast, 1819.
J. Pigot, *Commercial directory of Ireland . . . for 1820, 21 and 22*, Manchester, 1820.
William West, *A directory for Ireland, Scotland . . . for the years 1820, 1821, 1822*, [Cork?], 1820.
J. Pigot & Co., *Second edition of the commercial directory of Ireland . . . for 1821-22 and 1823*, Manchester, 1820.
Pigot & Co.'s City of Dublin and Hibernian provincial directory, London, 1824.
[Robert Donaldson], *Belfast directory for 1831-2*, Belfast, 1831.
William T. Matier, *Belfast directory 1835-6*, Belfast, 1835.
Matthew Martin, *Belfast directory for 1839*, Belfast, 1839.
Matthew Martin, *Belfast directory for 1840-1*, Belfast, 1840.
Matthew Martin, *Belfast directory for 1842-3*, Belfast, 1842.
Post Office Belfast annual directory, Belfast, 1843.
[John] Henderson's new Belfast directory and northern repository for 1843-44, Belfast, 1843.
Alexander Aldwell, *The City and County of Cork Post-Office general directory, 1844-5*, Cork, [1844].
Slater's national commercial directory of Ireland, Manchester, 1846.
[John] Henderson's Belfast directory and northern repository for 1846-47, Belfast, 1846.
[John] Henderson's Belfast directory and northern repository, Belfast, 1850.
[John] Henderson's Belfast directory and northern repository, Belfast, 1852.
[James Henderson], *Belfast and Province of Ulster directory for 1852, vol. I*, Belfast, 1852.

[James Henderson], *Belfast and Province of Ulster directory for 1854, vol. II*, Belfast, 1854.
Slater's (late Pigot & Co.'s) Royal national commercial directory of Ireland, Manchester and London, 1856.
[James Henderson], *Belfast and Province of Ulster directory for 1856, vol. III*, Belfast, 1856.
[James Henderson], *Belfast and Province of Ulster directory for 1858-9 vol. IV*, Belfast, 1858.
[Hugh] Adair's Belfast directory for 1860-1, Belfast, 1860.
[J.A. Henderson], *The Belfast and Province of Ulster directory for 1861-62, vol. V*, Belfast, 1861.
[J.A. Henderson], *The Belfast and Province of Ulster directory for 1863-64, vol VI*, Belfast, 1863.
Robert H. Laing, *Cork mercantile directory for 1863*, Cork, 1863.
[J.A. Henderson], *Belfast and Province of Ulster directory for 1865-66, vol. VII*, Belfast, 1865.
[R.W. Wynne], *Business directory of Belfast and principal towns in the Province of Ulster for 1865-66*, Belfast, 1865.
Henry and Coughlan's general directory of Cork for 1867, Cork, 1867.
[J.A. Henderson], *Belfast and Province of Ulster directory for 1868, vol. VIII*, Belfast, 1868.
[Benjamin Harvey], *The Belfast and Province of Ulster Post Office directory, vol. I*, Belfast, 1870.
Slater's (late Pigot & Co.'s) Royal national commercial directory of Ireland, Manchester and London, 1870.
The City of Cork directory . . . 1871, Liverpool, [1871].
Wilkie's Cork City directory . . . 1871, Liverpool, [1871].
Francis Guy's County and City of Cork directory for the years 1875 and 1876, Cork, [1875], and subsequent annual issues; the title varies slightly.
["The Belfast News-Letter"], *The Belfast and Province of Ulster directory for 1877, vol. IX*, Belfast, 1877. All subsequent volumes, from X to XLII, have been seen. They are for the years 1880, 1884, 1887, 1890, 1892, and annually from 1894.
[Brookes & Thompson], *The Belfast and District directory and Ulster guide for 1878*, London and Belfast, 1878.
Slater's Royal national commercial directory of Ireland, Manchester, 1881.
Purcell's commercial Cork almanac for the year of our Lord 1882, Cork, [1882?], and subsequent annual issues. The copy for 1890 has not been examined.

Provincial instrument makers

BARRY, – (Cork)

c. 1780	Known from a circumferentor[1].
	Joseph Barry was described as a watchmaker in 1824[2].

BENNETT, Thomas (Cork)

1810	Optician and Mathematical Instrument Maker, Patrick Street.
1812	Optician and Mathematical Instrument Maker, Patrick Street.
1812	Mathematical, Optical and Philosophical Instrument Maker, Patrick Street (Connor).
1817	Optician, Patrick Street.
1820	Optician, 2 Patrick Street (Pigot and Connor).
1824	Optician, 65 Patrick Street.
1826	Optician, 45 Patrick Street.
1828	Optician &c., 45 Patrick Street.
1844	Optician, 124 Patrick Street.
1846	Optician &c., 124 Patrick Street.

1856	Optician and Mathematical Instrument Maker, 124 Patrick Street.
1863	Optician, 124 Patrick Street.
1867	Optician, 124 Patrick Street.
	Taken over by Reynolds & Wiggins.

BENNETT, Thomas (Queenstown)

1863	15 Scott's Square.
	Known from an advertisement[3].

BINDA, Giovanni or John (Cork)

1846	Optician, 25 Duncan Street.
1852	Exhibited at Cork Exhibition.

BONCHER, William (Waterford)

fl. 1809	Mathematical Instrument Maker.
	Referred to by Westropp[4].

BOOKER, John (Cork)

1773	Looking-glass Seller (from Dublin). Opposite St Peter's Church, Main Street.
	See Chapter Five; traded chiefly in Dublin.

BOWMAN, Johnston (Belfast)

1870	Watchmaker, Jeweller and Optician, 52 High Street.
1877	Watchmaker, Jeweller and Optician, 82 High Street.
1878	Watch and Clockmaker, Jeweller etc., 82 High Street.

BURKE, William (Belfast)

1846	Spectacle Maker, 190 North Street.
1850-1854	Spectacle Maker and Book Dealer, 148 Millfield.
1856-1863	Spectacle Maker, 160 Millfield.

CAHOON BROTHERS (Belfast)

1877-1880	Watchmakers, Jewellers, Opticians, Bog Oak Carvers and Seal Engravers, 16 Castle Place.
1884	Jewellers, Silversmiths and Opticians, 16 and 18 Castle Place.
1887-1902	Watchmakers, Jewellers, Silversmiths and Opticians, 16 and 18 Castle Place.
1903-1906	Watchmakers, Jewellers, Silversmiths and Opticians, 33 Donegall Place.
1907-1908	Watchmakers, Jewellers, Silversmiths and Opticians, 27a Donegall Place.
1909-1911	Watchmakers, Jewellers, Silversmiths and Opticians, 33 Donegall Place and workshop, 27a Donegall Place.
	Succeeded by Weir & Sons (also at Grafton Street, Dublin).

CAPO, Anthony (Belfast)

1810	Barometer Maker, 154 Millfield.
1820	Barometer and Mirror Maker, 154 Millfield.
	Did not appear in directories for 1824, 1831 or 1835.

CAPPO, Anthony (Belfast)

1839-1840	Weatherglass Maker, 164 Millfield.
1843-1846	Barometer and Thermometer Manufacturer, 168 Millfield.
1850	Weatherglass, Thermometer etc. Manufacturer, 160 Millfield.
1852-1854	Optician, 160 Millfield.

CAPPO, Joseph (Belfast)

1835	[No trade], 168 Millfield.

1839	Weatherglass Maker, 202 North Street.
1840	Weatherglass Maker, 164 Millfield.
1856	Optician, 147 Millfield.
1858	Chemical and Philosophical Instrument Maker, 24 Portland Street.
1860-1861	Chemical and Philosophical Instrument Maker, 14 Portland Street.
1863-1865	Chemical and Philosophical Instrument Maker, 5 Portland Place.
1868	Philosophical and Chemical Instrument Maker, 5 Portland Place.
1870	Philosophical and Chemical Instrument Maker and Glass Blower, 5 Portland Place.
1877-1880	Philosophical and Chemical Instrument Maker, 5 Portland Place.

Cappo did not appear in street directories between 1842 and 1856; however, Bryden records a Joseph Cappo, hydrometer and philosophical instrument maker in Glasgow between 1847 and 1853[5].

CAPPO, Mary (Belfast)

1884	Philosophical Instrument Maker, 5 Portland Place.

CARR, Andrew (Belfast)

1908-1913	Watchmaker, Jeweller and Optician, 23 Cornmarket.

Succeeded by R. McDowell & Co.

CROZIER, James (Belfast)

1868	Watchmaker, Jeweller and Optician, 95 North Street.
1870	Watchmaker, Jeweller and Optician, 89 North Street.

CUMINE, James A. (Belfast)

1852-1856	Electrical and Optical Warehouse, 168 North Street.
1858	Philosophical Instrument Warehouse, 168 North Street. Philosophical Instrument Maker and Photographic Artist, 1 Joys Entry.
1860-1861	Optical and Photographic Warehouse, 168 North Street. Philosophical Instrument Maker and Photographic Artist, 1 Joys Entry.
1863	Philosophical Instrument Maker and Photographic Artist, 29 High Street.
1865	Dealer in Photographic Materials and Photographic Artist, 29 High Street.
1868	Photographic Artist and Photographic Material Merchant, 29 High Street.
1870	Photographic Artist and Philosophical Instrument Manufacturer, 27 High Street.
1877	Philosophical Instrument Maker and Photographic Establishment, 57 Upper Arthur Street.
1878	Philosophical Instrument Maker and Photographic Establishment, 48 Upper Arthur Street.
1880	Photographic Instrument Maker, 48 Upper Arthur Street.
1881	Optician and Philosophical Instrument Maker, 40-42 Upper Arthur Street.

CUTLER, Richard (Youghal)

1787	Watch and Clockmaker, Engraver and Barramatitian [i.e. Barometrician].

CUTTLER, Richard (Carrickfergus)

1782	Barometer Maker, Watchmaker & Engraver, Castle Gate.

From a newspaper advertisement[6]. Probably the same man as Richard Cutler of Youghal.

FITTON, Alexander (Cork)

1787	Philosophical, Optical and Mathematical Instrument Maker, 60 Paul Street.

FITTON, Francis (Cork)

1783	Instrument Maker. Voted at election[7].
1784	Gratton Street, Cork. In the Peabody Museum, Salem, there is an octant marked "Made by Francis Fitton Cork For Captn Abel Orpin 1784".
1787	Dealer in Glass and Earthenware, Ship-chandler, and Mathematical Instrument Maker, 36 Paul Street, corner of Half Moon Street.
1795	Instrument Maker, Paul Street.

FITTON, William (Cork)

fl. 1773	Known from an octant, signed "Made by William Fitton Corke for Thomas Hickman Esq 1773"[8].

FITZGERALD, Michael (Cork)

1820	Land Surveyor and Mathematical Instrument Maker, Sullivan's Quay (Pigot).
1820	Land Surveyor, Sullivan's Quay (Connor).

GARDNER, Henry (Belfast)

1809	Watchmaker and Dentist, successor to J. Rider, 27 High Street.
1819	Watch and Clockmaker, Silversmith and Jeweller, Optical and Mathematical Instrument Maker, 65 High Street.
1820	Jeweller, Gold and Silversmith and Dealer in Watches, Money Exchange Broker, 65 High Street.
1824	Watchmaker and Optician, 57 High Street.
1831	Watchmaker and Jeweller, 57 High Street.
1835	Watchmaker, Jeweller and Optician, 57 High Street. In 1809, Gardner was given as "Henry L." G. Fennell gives "Gardner, R.I." In partnership with Job Rider 1805-1807, with R. Neill 1809-1818. Watch inscribed "Gardner and Neill" in Victoria and Albert Museum, South Kensington[9]; microscope signed "Gardner & Neal, Belfast", Sotheby's 20 May 1974 Lot 21; 10-inch octant which once belonged to Professor James Thomson, father of Lord Kelvin, signed "Gardner & Neil * Belfast", now in the Natural Philosophy Collection, University of Glasgow.

GEARY, Francis (Cork)

1805	Instrument Maker, Paul Street.
1810	Optician, Paul Street.
1812	Optician, Paul Street.
1812	Optician and Instrument Maker, Paul Street (Connor).
1817	Optician, Paul Street.
1820	Mathematical Instrument Maker, Lapps Island (Connor).

GIBSON, William (Belfast)

1868	Watchmaker, Jeweller, Silversmith and Optician, 5 Castle Buildings and 174 North Street.
1870	Watchmaker, Jeweller, Silversmith, Optician and Bog Oak Carver, 36 Castle Buildings.
1877	Watchmaker, Jeweller, Silversmith, Optician and Manufacturer of Bog Oak Jewellery, Donegall Place.
1878	Watchmaker and Jeweller, Donegall Place and 34 and 36 Castle Place.

GIBSON & Co. (Belfast)

1880-1890	Watchmakers, Jewellers, Silversmiths, Opticians and Manufacturers of Bog Oak Jewellery, Donegall Place and 34 and 36 Castle Place.
1892	Watchmakers, Jewellers, Silversmiths and Opticians, Donegall Place and 34 and 36 Castle Place.

GIBSON & Co. Ltd. (Belfast)

1894-1921	Watchmakers, Jewellers, Silversmiths and Opticians, Donegall Place and 34 and 36 Castle Place.

GRIBBEN, Edward (Belfast)

1840	Watchmaker, Silversmith, Jeweller and Optician, 84 High Street.
1868	Watchmaker, Jeweller, Optician and Silversmith, 13 High Street.
1870	Watchmaker, Jeweller, Optician and Silversmith, 7 High Street.
	Gribben first appeared in the directories in 1831 as a watchmaker and jeweller, and made his final appearance in 1892 as a watchmaker; on only three occasions – 1840, 1868, and 1870 – was he described as "optician".

HAMILTON, Robert H. (Belfast)

1892	Watchmaker, Jeweller and Optician, 30 Church Lane.
1894-1900	Watchmaker and Optician, 30 Church Lane.
	Hamilton appeared in the directories for 1884, 1887 and 1890 as a watchmaker and jeweller at 30 Church Lane.

HAMILTON, Samuel (Belfast)

1880	Watchmaker, Jeweller and Optician, 13 Cromac Street.
	Hamilton appeared in the directories for 1877 and 1878 as a watchmaker and jeweller at 13 Cromac Street.

HARBINSON, J.F. & Co. (Belfast)

1868	Jewellers, Watchmakers and Opticians, 70 High Street and corner Victoria Street.
1870-1877	Jewellers, Watchmakers etc., 85 High Street.
1878	Watch Manufacturers, Jewellers, Silversmiths, Opticians, Bog Oak Carvers etc., 85 High Street.
1880	Jewellers, Watchmakers etc., 85 High Street.
	John F. Harbinson began as a pawnbroker at 1 Cranston Place in 1858, and appears as a jeweller in the directories for 1860 and 1861; J.F. Harbinson & Co. appear as watchmakers, jewellers etc. in the directories for 1863 and 1865.

HARBINSON, J.F. & Son (Belfast)

1884	Watch and Clock Manufacturers, Jewellers, Opticians etc., 1 Royal Avenue.
1887	Jewellers and Opticians, 1 Royal Avenue.

HUNT, Charles V. (Belfast)

1887	Optician, 63A Queen Ann's Buildings, Ann Street.
1890	Oculist and Optician, 63 Ann Street.
1894-1904	Occulist and Optician, 63 Ann Street.
	Hunt did not appear in the directory for 1892.

HUNT & Son (Belfast)

1892	Oculist Opticians and Nautical Instrument Makers, 63 Ann Street.

HUNT, Henry (Cork)
- 1844 Optician, 118 Patrick Street.
- 1846 Optician and Mathematical Instrument Maker, 118 Patrick Street.
- 1881 Optician, 109 Old George Street.
- 1883 Optician, 109 George's Street.
- 1884 Optician, 109 George's Street.

HUNT, James (Cork)
- 1817 Optician, Drawbridge Street.

HUNT, John (Cork)
- 1817 Optician, Patrick Street.

HUNT, Thomas I (Cork)
- 1792 Said to have started trading.
- 1805 Mathematical Instrument Maker, Patrick Street.
- 1810 Optician, Patrick Street.
- 1812 Optician, Patrick Street.
- 1812 Optician, Blackrock Road (Connor).

HUNT, Thomas II (Cork)
- 1820 Optician, 37 Patrick Street (Connor and Pigot).
- 1824 Optician, 65 Patrick Street.
- 1826 Optician and Mathematical Instrument Maker, 20 Patrick Street.
- 1828 Optician and Mathematical Instrument Maker, 20 Patrick Street.
  The 1820 entry in Pigot is for "Hunt Thomas, jun."

HUNT & Son (Cork)
- 1886-1892 Opticians, 109 George's Street.
- 1893-1895 Opticians, 109 George's Street and 6 Patrick Street.

JOHNSTON, J.W. (Belfast)
- 1917-1920 Optician and Jeweller, 204 Duncairn Gardens.

KNIE, Balthazar (Cork)
- 1773 Mathematical Instrument Maker.
  Referred to by Westropp[10]. Knie settled in Edinburgh in 1776[11]. A letter was sent to the editor of a Cork newspaper in 1773, about the behaviour of barometers, signed "Balthasar Weatherwise"[12].

LAMONT, Carson (Belfast)
- 1846 Watch and Clockmaker and Optician, 15 Castle Place.

LAMONT & Co. (Belfast)
- 1843 Opticians, Watch and Clockmakers, Jewellers and Marine Instrument Repairers, 15 Castle Place.
- 1846 Watchmakers, Jewellers, Silversmiths and Opticians, 15 Castle Place.

LAMONT, John jun. (Belfast)
- 1842 Watchmaker, Jeweller, Silversmith and Optician, 15 Castle Place.

LEE, Joseph (Belfast)
- 1835 [no trade], 28 Gloucester Street.
- 1839 Watchmaker, Jeweller, Optician etc., 24 High Street.
- 1840-1842 Watchmaker, Jeweller, Optician etc., 74 High Street.

1843	Watchmaker, Jeweller, Optician and Silversmith, 74 High Street.
1846	Watchmaker, Jeweller, Optician and Silversmith, 57 High Street.
1868	Jeweller to the Queen, Watchmaker, Goldsmith and Optician, 57 High Street.
	Succeeded by Lee & Son.

LEE & Son (Belfast)

1850	Jewellers, Silversmiths, Watchmakers and Opticians, 57 High Street.
1852-1854	Jewellers to the Queen, Watchmakers, Silversmiths and Opticians, 57 High Street.
1856	Jewellers, Watchmakers and Opticians, 57 High Street.
1858-1861	Watchmakers, Silversmiths and Opticians, 57 High Street.
1863	Jewellers to the Queen, Watchmakers, Silversmiths and Opticians, 57 High Street.
1865	Watchmakers, Jewellers and Silversmiths, 57 High Street.
1870	Jewellers to the Queen, Watchmakers, Silversmiths and Opticians, 57 High Street.

LE GROS, Peter Joseph (Belfast)

1839	Watch, Clock and Mathematical Instrument Maker, 46 Joy Street.
	Le Gros appeared in the directory for 1835 as a watch and clockmaker at 22 Joy Street.

LIZARS, J. (Belfast)

1894-1900	Manufacturing Optician and Dealer in Photographic Apparatus, 73 Victoria Street.
1907-1921	Manufacturing Optician and Dealer in Photographic Apparatus, 8 Wellington Place.
	Branch of the Glasgow-based firm[13].

LOWRY, John (Belfast)

1850	Watch and Chronometer Maker, 55 High Street.
1852-1856	Watch and Chronometer Maker, 19 High Street.
1863-1865	Jeweller and Watchmaker, 66 High Street.
1868	Jeweller, Chronometer, Watch and Clockmaker, 66 High Street.
1870	Jeweller, Chronometer, Watch and Clockmaker, 46 High Street.
1877	Practical Watch and Clockmaker etc., 46 High Street.
1880	Chronometer, Watch and Clockmaker etc., 94 High Street.
	In the directories for 1858, 1860 and 1861, the firm appeared as J. & S. Lowry.

LOWRY, J. & S. (Belfast)

1858-1861	Jewellers and Watchmakers, 66 High Street.

LUKE, Edward (Belfast)

1868	Watchmaker, Jeweller and Optician, 23 North Street.
	Luke appeared at the same address in the directory for 1865, as a watchmaker and jeweller.

McCLEAVE BROTHERS (Belfast)

1868	Jewellers, Watchmakers, Opticians, and Sextant and Quadrant Regulators, 59 Donegall Street.
1870	Jewellers, Watchmakers, Opticians, and Sextant and Quadrant Regulators,

73 Donegall Street.
McCleave Brothers first appeared in the 1865 directory as watchmakers, jewellers etc. at 7 Ann Street, and made a final appearance in 1877 as watchmakers, jewellers, stone seal engravers etc. at 73 Donegall Street.

McCONNELL, John & J. (Belfast)
1892 Watch and Clockmakers, Jewellers and Opticians, 74 North Street.
John McConnell first appeared as a jeweller at 265 York Street in the 1877 directory; John & J. McConnell continued as watchmakers and jewellers from 1894 until 1908; from 1909 until their last appearance in 1916 they were listed as diamond dealers, watchmakers and jewellers.

McCORMICK, Hugh (Belfast)
1858 Manager for Neill Brothers.
1860-1861 Salesman in Neill Brothers.
1865 Watchmaker, Jeweller etc., 5 Castle Buildings.

McCORMICK & MEARNS (Belfast)
1863 Watchmakers and Jewellers, 5 Castle Buildings.

McCUTCHEON & DONALDSON (Belfast)
1901-1905 Watch and Clockmakers, Opticians, Engravers, Watch and Clock Tool and Material Dealers, Wholesale Importers of English, American and French Clocks and Bronzes, 113 Royal Avenue.
1906-1910 Watch and Clockmakers and Jewellers, 113 Royal Avenue.
1911-1916 Watch and Clockmakers and Jewellers, 113A Royal Avenue.
1917-1920 Watch and Clockmakers and Jewellers, 113A Donegall Road.
1921 Watch and Clockmakers and Jewellers, 7 North Street.
Successors to Steel & Son.

McDOWELL, R. & Co. (Belfast)
1911-1913 Watchmakers, Jewellers, Gold and Silversmiths and Opticians, 14 High Street.
1914-1916 Watchmakers, Jewellers, Gold and Silversmiths and Opticians, 14 High Street and 23 Cornmarket.
1917 Watchmakers, Jewellers, Gold and Silversmiths and Opticians, 14 High Street and 68 Ann Street.
1918-1921 Watchmakers, Jewellers, Gold and Silversmiths and Opticians, 14 High Street, and 68, 70 Ann Street.

MacHUGH, Mrs Margaret (Cork)
1863 Optician, 78 George's Street.
1867 Optician, 78 George's Street.
1870 Optician, 78 George's Street.
1871 Optician and Spectacle Maker, 78 George's Street.
1875 Optician, 78 George's Street.
1883 Optician, 78 George's Street.
Although she appears in Guy for 1883, Mrs MacHugh is not in Slater for 1881, and Purcell for 1882 and 1883 gives Thomas Quirke, optician, at her address.

MASON, Jonathan (Limerick)
1846 Optician, 6 Patrick Street.
Previously in business in Dublin with his brother Thomas between 1805 and

1808, and 1813 and 1817; and on his own from 1809 to 1812, and from 1818 to 1822. He moved to Limerick after his second marriage, where he carried on an optical business until his death in 1849[14].

MEARNS & Co. (Belfast)

1868	Watch Manufacturers, Goldsmiths, Opticians etc., 24 High Street.
1870	Watch Manufacturers, Goldsmiths, Opticians etc., 16 High Street.
	John Mearns appears in the directory for 1865 as a watchmaker, jeweller and silversmith at 39 High Street; see McCormick & Mearns.

MINNIECE, Jas. A. (Londonderry)

1865	Optician and Jeweller, 10 Shipquay Street.
1868-1870	Watchmaker, Jeweller and Nautical Instrument Regulator, Shipquay Street.
1877	Watch and Clockmaker, Shipquay Street.
1878	Goldsmith, Jeweller, Watchmaker and Optician, 23 Shipquay Street.
1880-1913	Watchmaker and Clockmaker, Shipquay Street.
	Successor to Neills & Minniece.

MITCHELL, Richard (Cork)

1773	Mathematical Instrument Maker.
	Referred to by Westropp[15].

MOORE, F.M. (Belfast)

1854-1858	Chronometer and Watchmaker, 114 High Street.
1860	Chronometer and Practical Watchmaker, 114 High Street.
1861-1863	Chronometer, Watchmaker and Optician, 114 High Street.
1865	Watchmaker, Jeweller etc., 114 High Street and 23 Edenquay, Dublin.
1868	Chronometer and Practical Watchmaker, Optician etc., 114 High Street.
1870	Chronometer and Practical Watchmaker, Optician etc., 102 High Street.
1877-1880	Chronometer and Practical Watchmaker, Nautical Instrument Maker, Optician etc., 102 High Street and 23 Edenquay, Dublin.
1887	Chronometer and Watchmaker and Optician, 102 High Street.
1890-1898	Practical Chronometer and Watchmaker and Optician, 100 High Street; 23 Edenquay, Dublin.

MOORE, F.M. Ltd. (Belfast)

1899-1905	Chronometer and Practical Watchmaker, Nautical Instrument Maker, Optician etc., 102 High Street; 23 Edenquay, Dublin.
1906-1906	Chronometer and Practical Watchmaker, Nautical Instrument Maker, Optician etc., 18 Donegall Place.

MYERS, George (Belfast)

1819	Barometer and Looking-glass Manufacturer, 104 High Street.
1820	Barometer and Mirror Maker, 104 High Street.

NEILL BROTHERS (Belfast)

1850	Watchmakers, Jewellers, Silversmiths, Opticians etc., Wholesale and Retail, 23 High Street.
1852-1854	Watchmakers, Jewellers, Silversmith and Opticians, 23 High Street.
1856	Watchmakers, Jewellers and Opticians, 23 High Street.
1858	Jewellers and Watchmakers, 23 High Street.
1860-1861	Watch and Chronometer Makers, Jewellers, Silversmiths and Opticians, 23 High Street.

1863 Goldsmiths, Silversmiths, Watch and Chronometer Makers and Opticians, 23 High Street.
Succeeded by James Neill & Co., and John R. Neill.

NEILL, James (Belfast)
1856-1861 Watchmaker, Jeweller and Optician, 6 Donegall Place.
In 1854 Neill was listed as a watchmaker at the same address.

NEILL, James & Co. (Belfast)
1865 Watchmakers, Jewellers, Silversmiths etc., 8 Donegall Place.
1868 Goldsmiths, Silversmiths, Watch and Chronometer Makers, Jewellers and Opticians, 8 Donegall Place.
1870-1877 Goldsmiths, Silversmiths, Watch and Chronometer Makers, Jewellers, Opticians and Bog Oak Workers, 14 Donegall Place.
1878 Watchmakers, Jewellers, Silversmiths, Opticians, and Bog Oak Carvers, 14 Donegall Place and 33 Castle Lane.
1880 Goldsmiths, Jewellers, Watchmakers, Silversmiths and Opticians: Manufacturers of Bog Oak Jewellery, 14 Donegall Place.

NEILL, John R. (Belfast)
1865 Watchmaker, Jeweller, Silversmith etc., 23 High Street.
1868 Watchmaker, Jeweller, Silversmith and Optician, 23 High Street.
1870-1877 Watch, Chronometer and Clockmaker, Jeweller, Silversmith and Optician; Agent for the U.K. Provident Institution, 21 High Street.
1878 Watchmaker, Jeweller, Silversmith and Optician, 21 High Street.
1880 Watch, Chronometer and Clockmaker, Jeweller, Silversmith and Optician; Agent for the U.K. Temperance Institution, 21 High Street.

NEILL, Robert (Belfast)
1805 Clock and Watchmaker, High Street.
1809 Watchmaker, High Street.
1819-1820 Watch and Clockmaker, 25 High Street.
1824 Watch and Clockmaker, 21 High Street.
1831 Watchmaker and Jeweller, 25 High Street.
1835-1840 Watch and Clockmaker, Jeweller, Silversmith and Optician, 25 High Street.
Continued as Robert Neill & Sons.

NEILL, Robert & Sons (Belfast)
1842-1846 Watch and Clockmakers, Jewellers, Silversmiths and Opticians, 25 High Street.
From 1843 John R. Neill of R. Neill & Sons is listed. Succeeded by Neill Brothers.

NEILL, Robert & Sons (Londonderry)
1843 Watch and Clock Manufacturers, Gold and Silversmiths, Opticians, Nautical Instrument, Globe and Chart Sellers, 6 Diamond.
1846 Watch and Clock Manufacturers, Gold and Silversmiths, Opticians, Nautical Instrument, Globe and Chart Sellers, 6 Diamond.
1846 Watch and Clockmakers and Silversmiths, Jewellers and Opticians, 27 Shipquay Street (Slater).

NEILL, Sharman D. (Belfast)
1884 Watchmaker, Jeweller etc., 12 Donegall Place.

1887	Jeweller and Diamond Merchant, Watch and Clockmaker, 12 Donegall Place.
1890	Jeweller, Silversmith, Optician, Watch and Clock Manufacturer, by Special Appointment to HRH Prince of Wales, 12 Donegall Place.
1892	Jeweller, Silversmith, Watch and Clock Manufacturer, 12 Donegall Place.
1894-1897	Jeweller and Silversmith, Watch and Clock Manufacturer; Turret Clocks a Speciality (by appt. to HRH the Prince of Wales), 12 Donegall Place.
1898-1901	Jeweller and Silversmith, Watch and Clock Manufacturing, 12 Donegall Place.
1902-1903	By appointment to His Majesty the King, Jeweller and Silversmith, Watch and Clock Manufacturing, 12 Donegall Place.
1904-1906	By appointment to His Majesty the King, Jeweller and Silversmith, Watch and Clock Manufacturing, Tower Clocks a Speciality, 12 Donegall Place.
1907-1909	Jeweller and Silversmith, Watch and Clockmaker; Turret Clocks a speciality. By appointment to His Majesty the King, 22 Donegall Place.

NEILL, Sharman D., Ltd (Belfast)

1910-1921	Jeweller and Silversmith, Watch and Clockmaker; Turret Clocks a Speciality. By appointment to His Majesty the King, 22 Donegall Place.

NEILLS & MINNIECE (Londonderry)

1852-1861	Watchmakers, Jewellers and Opticians, Shipquay Street. Successor to Robert Neill & Sons; succeeded in 1865 by Jas A. Minniece. The firm does not appear in the 1863 directory.

NICHOLL, S. (Belfast)

1901-1904	Watchmaker, Jeweller, Silversmith and Optician, 35 Bridge Street.
1905-1907	Watchmaker, Jeweller, Silversmith and Optician, 33 Bridge Street.
1908	Watchmaker, Jeweller, Silversmith and Optician, 35 Bridge Street.

NOSEDA, John (Belfast)

1774	Weatherglass maker. Died April 1779[16].

PERCIVAL, Gerald (Cork)

1881	Electrical and Philosophical Instrument Maker, 8 Robert Street.
1882-1883	Engraver and Electrical Instrument Maker, 8 Robert Street.

PRINCE, Abraham (Waterford)

1839	Optician and Mathematical Instrument Maker. Referred to by Westropp[17].

RANKIN Bros. (Belfast)

1877	Watchmakers, Jewellers and Opticians, 5 Castle Lane.
1878	Watch and Clockmakers, Jewellers and Opticians, 5 Castle Lane.
1880	Jewellers etc., 5 Castle Lane.
1887	Watchmakers, Jewellers etc., 5 Castle Lane.
1890-1913	Watchmakers, Jewellers and Opticians, 17 Royal Avenue.
1914-1916	Watchmakers, Jewellers and Opticians, 17 Royal Avenue and 115 Albert-bridge Road. In 1880 the directory names the brothers as Hugh Rankin, Sullivan Place, Holywood, and Alex. Rankin of Cairo, Egypt. In 1908, J.H.P. Rankin, Fellow of the Institute of Opticians, was associated with the firm.

RANKIN Bros. (Belfast) Ltd (Belfast)

1917-1920 Watchmakers, Jewellers and Opticians, 17 Royal Avenue and 115 Albertbridge Road.

1921 Watchmakers, Jewellers and Opticians, 17 Royal Avenue.

REYNOLDS & WIGGINS (Cork)

1870-1872 Opticians, 125 Patrick Street.
Successors to Thomas Bennett.

RIBOLDI, Joseph (Limerick)

1824 Carver and Gilder, 122 George's Street.
An 8-inch wheel barometer signed "L. Riboldi, Limerick" was sold at Phillips 8 May 1985 Lot 1. Goodison notes a wheel barometer and a stick barometer by "J. Ribaldi, Limerick"[18].

SILO, Modesto (Londonderry)

1843 Carver and Gilder, Looking-glass, Gilt Picture Frame, Room Bordering, Barometer and Thermometer Manufacturer, 21 Cornmarket (Post Office).

1846 Carver and Gilder, Looking-glass, Barometer and Thermometer Manufacturer, 21 Cornmarket (Slater).
The firm continued as carvers and gilders into the 1850s.

SIMPSON, W.J. (Belfast)

1890 Jeweller, Watchmaker, Goldsmith, Silversmith and Optician, 33 Donegall Street.

1892 Goldsmith and Jeweller, 4 Bridge Street.

1894-1902 Goldsmith and Jeweller, 10 Cornmarket.

STEEL & Sons (Belfast)

1877 Manufacturing Jewellers, Watch and Clockmakers, Opticians etc., 16 High Street.

1878 Watch and Clockmakers, Manufacturing and Masonic Jewellers, Opticians, Watchglass Manufacturers and Watchmakers' Tools and Material Dealers, 16 High Street.

1880 Jewellers, Watchmakers etc., 19 Arthur Street.

STEEL, M.C. & Sons (Belfast)

1884 The Irish Electro-gilding, Silver and Nickelplating, Electrotyping, Bronzing and Polishing Works, Watchmakers etc., 39 Ann Street.
Successors to Steel & Sons; the firm does not appear in the 1887 directory.

STEEL & Sons Ltd (Belfast)

1890 Clock and Watchmakers, Jewellers, Silversmiths, Bronze and Metal Polishers, 113 Royal Avenue.

1892-1900 Electro-plate Manufacturers, Gilders and Platers, Watchmakers and Jewellers, Gold and Silver Smiths, Engravers, Opticians, Brass and Metal Polishers and Lacquerers, Watch and Clock Tool and Material Dealers, Nickel Platers and Bronzers, Watchglass Makers, Wholesale French and American Clock Importers, 113 Royal Avenue.
Succeeded by McCutcheon & Donaldson.

SWEENY, Edward (Cork)

1763 Instrument Maker[19].
Married Mary Green, 1772, who may be "widow Sweeny" below[20].

SWEENY, Edward Nathaniel (Cork)
1798 Mathematical and Optical Instrument Maker, 27 Patrick Street.
Known from an advertisement[21].

SWEENY, Nathaniel (Cork)
fl. 1702 Instrument Maker.
Described as "deceased", father of Edward Sweeny[22].

SWEENY, Widow (Cork)
1795 Instrument Maker, Paul Street.
See Edward Sweeny, above.

VOSTER, Ann (Cork)
1760 Known from an advertisement[23].

VOSTER, Elias (Cork)
d. 1760 Instrument Maker.
Described as "the late Mr Elias Voster"[24].

VOSTER, Daniel (Cork)
1742-1760 Instrument Maker.
The Science Museum, London, has a slide rule signed "Daniel Voster Corke Fecit 1742"[25]. He died in May 1760. A sundial, signed "D Voster Corke Fe 1751" is in a private Collection.

WALLACE, Richard (Limerick)
1856 Jeweller, Watchmaker and Optician, 125 George Street.
1870 Watchmaker, Jeweller and Optician, 129 George Street.
1881 Watchmaker and Optician, 129 George Street.

WALSH, Michael (Cork)
1812 Instrument Maker, corner of Fair Lane (Connor).

WEIR, Robert (Coleraine)
1846 Watch and Clockmaker, and Jeweller, Optician, etc., 17 Bridge Street.

WEIR & Sons (Belfast)
1912-1920 Jewellers, Watch and Clock Manufacturers, Goldsmiths, Silversmiths and Opticians. Also Grafton Street, Dublin, 33 Donegall Place.
Successors to Cahoon Bros. The firm was described in Thom's *Dublin Directory* as "jewellers and silversmiths".

WIGHTMAN, David (Belfast)
1880-1884 Jeweller, 8 Castle Street.
1887 Watchmaker and Manufacturing Jeweller, 8 Castle Street.
1890 Manufacturing Optician, Watchmaker, Jeweller, Electroplater, Bog Oak Carver and Clockmaker, 4, 6 and 8 Castle Street.

WIGHTMAN, D., Ltd. (Belfast)
1892 Manufacturing Jewellers, Opticians, Watch and Clockmakers, 20 Royal Avenue.
1894 Jewellers, Watch and Clockmakers, 20 Royal Avenue.

WIGHTMAN & Co. (Belfast)
1895-1898 Watchmakers and Jewellers, 19 Garfield Street.

1899-1900	Watchmakers and Jewellers, 5 Garfield Street.

WIGHTMAN, D. (Belfast)

1901-1904	Jeweller, 188 North Street.

WILLIAMS, Joseph (Londonderry)

1865	Jeweller, 6 Shipquay Street.
1868-1870	Watchmaker, Jeweller and Regulator of Nautical Instruments etc., Shipquay Street.
1877-1890	Watch and Clockmaker, Shipquay Street.

WILSON, James (Belfast)

1852-1856	Watchmaker, Jeweller and Optician, 7 High Street.
1858-1863	Watchmaker and Jeweller, 7 High Street.
1865-1868	Watchmaker and Jeweller, 103 Victoria Street.

WOODS, Michael (Belfast)

1868	Jeweller, 25 Spencer Street.
1870	Watchmaker, Jeweller and Optician, 11 Donegall Street.
1877-1880	Jeweller, Watchmaker, 283 York Street.

References and Notes for Chapter Eight

1. *Bulletin of the Scientific Instrument Society*, no. 1, 1983, 7.
2. Geraldine Fennell, *A list of Irish watch and clock makers*, Dublin, 1963, 3.
3. Advertisement in Robert H. Laing, *Cork mercantile directory for 1863*, Cork, 1963, 27.
4. M.D. Westropp, "List of Dublin opticians and mathematical instrument makers", manuscript at the Museum of the History of Science, Oxford.
5. D.J. Bryden, *Scottish scientific instrument-makers 1660-1900*, Edinburgh, 1972, 46.
6. *Belfast News-Letter*, 19 March 1782.
7. *Exact list of the freemen and freeholders who voted at the election . . . city of Cork . . . 1783*, Cork, 1783, 13.
8. Information from Dr Charles Mollan, and also from Peter Delehar.
9. Fennell, *op. cit.*, 15 (ref. 2).
10 Westropp, *op. cit.*, (ref. 4).
11. Knie is discussed in Chapter 5.
12. *Hibernian Chronicle*, 18 January 1773.
13. For more on the firm of Lizars, see T.N. Clarke, A.D. Morrison-Low and A.D.C. Simpson, *Brass & glass: Scottish scientific instrument making workshops in Scotland as illustrated by instruments from the Frank collection at the Royal Museum of Scotland*, Edinburgh, 1989.
14. [Thomas H. Mason & Sons Ltd], *The Mason family business: a brief history 1780-1980*, [Dublin, 1980], 4.
15. Westropp, *op. cit.*, (ref. 4).
16. *Belfast News-Letter*, 27 April 1779. He advertised in the same newspaper 12-16 August 1774. See also, Fennell, *op. cit.*, 29 (ref. 2), who gives his name as "NOSEDON (NOREDON?), John".
17. Westropp, *op. cit.*, (ref. 4).
18. Goodison, *op. cit.*, 352 (ref. 11).
19. J.H. Andrews, *Plantation acres: an historical study of the Irish land surveyor*, Belfast, 1985, 263, n.92 quotes the *Cork Evening Post*, 6 October 1763. This stated: "EDWARD SWEENY, son of the late Nathaniel Sweeny, deceased, begs leave to inform the public, that he continues to make all kinds of Mathematical Instruments, in the same correct manner as his late father did, and hopes by his care and diligence to merit the favour and esteem of all such as are pleased to deal with him. N.B. All kinds of Copper plate Printing are done in the neatest manner, and at reasonable rates by said Swiney [sic]".
20. *Hibernian Journal*, 7 October 1772.
21. *The Harp of Erin*, 10 March 1798.
22. *Cork Evening Post*, 6 October 1763.
23. *Cork Evening Post*, 16 June 1760.
24. Andrews, *op. cit.*, 263, n.92 (ref. 19), quotes *Cork Evening Post*, 16 June 1760.
25. Science Museum, inventory no. 1907-108: the item was purchased from J. Lecky, who was also the source of the Valencia Island mariner's astrolabe. Nothing further is known about the history of the rule.

INDEX